Cinematic Algorithms

The Rise of Generative AI in Video Art and Visual Culture

James Hutson
Professor of Art History, AI, and Visual Culture
Lindenwood University
USA

Andrew Allen Smith
Associate Professor of Social Media Marketing
Lindenwood University
USA

CRC Press
Taylor & Francis Group
Boca Raton London New York

CRC Press is an imprint of the
Taylor & Francis Group, an **informa** business
A SCIENCE PUBLISHERS BOOK

Cover credit: Cover design created by James Hutson using DALLE-3.

First edition published 2025
by CRC Press
2385 NW Executive Center Drive, Suite 320, Boca Raton FL 33431

and by CRC Press
4 Park Square, Milton Park, Abingdon, Oxon, OX14 4RN

CRC Press is an imprint of Taylor & Francis Group, LLC

Library of Congress Cataloging-in-Publication Data (applied for)

ISBN: 978-1-032-98777-4 (hbk)
ISBN: 978-1-032-98877-1 (pbk)
ISBN: 978-1-003-60112-8 (ebk)

DOI: 10.1201/9781003601128

Typeset in Times New Roman
by Prime Publishing Services

Dedication

This manuscript is dedicated to our families, whose unwavering support and encouragement have been the cornerstone of our journey. To our parents, we express deep gratitude for instilling in us a thirst for knowledge and the determination to pursue our passions. Your belief in our abilities and unwavering support have propelled us forward.

To our spouses and partners, we extend heartfelt appreciation for standing by our side throughout the long hours of research and writing. To our children, we are grateful for your inspiration and the joy you bring to our lives. Your presence reminds us of the importance of our work and motivates us to strive for excellence. To our extended families, friends, and loved ones, we thank you for your encouragement, words of wisdom, and unwavering belief in our abilities. In particular, James Hutson would like to thank his wife Piper and children Bishop and Aurora.

Your support has been a source of strength and inspiration. We acknowledge the sacrifices you have made and the understanding you have shown, allowing us the time and space to pursue our scholarly endeavors. This manuscript stands as a testament to your love and support. With deepest appreciation and heartfelt gratitude, we dedicate this work to our families.

Epigraph

"Technology, like art, is a soaring exercise of the human imagination."

— Daniel Bell

Preface

From the earliest days of film, the potential of technology to transform storytelling has been a driving force behind innovation. When Georges Méliès unveiled *Le Voyage dans la Lune (A Trip to the Moon)* in 1902, audiences marveled at his groundbreaking special effects. This short film, with its fantastical journey to the moon, brought visual effects to the forefront, showcasing how imaginative scenes and visuals could captivate audiences through the power of the emerging medium of cinema. As the film industry matured, techniques like stop motion and matte paintings allowed directors to create new worlds, such as the giant ape in *King Kong* (1933) or the fantastical landscapes in *The Wizard of Oz* (1939). These methods were revolutionary, giving storytellers tools to visualize stories that previously lived only in their imaginations.

Decades later, the advent of computer-generated imagery (CGI) brought about a new era. *Jurassic Park* astonished audiences in 1993 with its realistic dinosaurs, setting a high bar for special effects that could be as awe-inspiring as the narrative itself. This was followed by advancements in motion capture, famously utilized in films like *Avatar* (2009) and *The Lord of the Rings* (2001–2003), where the character Gollum set a new standard for CGI realism. The technological innovations of each era enabled filmmakers to tell stories that would have been impossible to convey otherwise, thus pushing the boundaries of cinematic creativity.

The latest generation of tools, powered by generative artificial intelligence (GAI), has introduced a shift that goes beyond merely enhancing visual effects: it democratizes the medium itself. Tasks that once required large teams of specialists can now be accomplished by a single creator with access to AI-powered platforms. A solitary filmmaker can, with these tools, fulfill the roles of director, cinematographer, editor, and visual effects artist, all from a personal computer. This shift in technology has not only widened access to high-quality production but has also diversified the backgrounds of those creating films. While traditional production centers like Los Angeles, Mumbai, Beijing, and Seoul have long dominated the industry, the new generation of AI creatives is dispersed across the globe, from rural Southern India to the French countryside.

As this book will explore, these filmmakers see the GAI not merely as a tool but as an empowering, popularizing force for storytelling. The technology allows them to bypass traditional gatekeeping structures within the industry. Without the necessity of expensive equipment, studio approval, or large budgets, a wider range of voices are now finding their way into the medium. This shift means that films can emerge that reflect a more diverse array of perspectives, stories, and cultural experiences than ever before. Many of the filmmakers featured in this book come from non-traditional backgrounds, ranging from marketing and programming to fine arts and photography. Despite their varied origins, they are united by a deep understanding of narrative and a desire to push the boundaries of visual storytelling.

The creative processes of these filmmakers vary greatly. Some start with generative image tools, like Midjourney, to produce initial concepts or storyboards, which then evolve into full video sequences. Others begin with their own photography or artwork, feeding these images into AI video generators like Pika or Runway to produce short clips. Some employ video-to-video techniques, transforming their own footage into entirely new creations. Once generated, these clips often undergo further refinement in editing software such as Adobe Premiere Pro, with additional enhancements added in visual effects software like Cinema 4D. The versatility of these workflows illustrates the flexibility that new GAI tools bring to filmmaking—allowing creators to work in unique and highly personalized ways that best suit their vision.

Yet, as with every major shift in the industry, the rise of AI-driven filmmaking brings uncertainty for established studios. While some, like Lionsgate, are already experimenting with AI for pre-visualization, others are more cautious. Industry leaders, including directors like James Cameron, have joined AI company boards, suggesting that the integration of the technology into the filmmaking process is not merely speculative but inevitable. Historically, every technological leap—from CGI to digital editing—has reshaped the landscape, creating new roles while rendering others obsolete. The rapid evolution of AI stands poised to effect similar changes but at an unprecedented pace.

AI filmmaking, while rooted in a long lineage of cinematic innovation, is unfolding in an era of faster and more disruptive change. In the past, as CGI became more prevalent, stop-motion artists and matte painters saw their roles diminish, replaced by new positions in visual effects. Now, the pace of advancement means that filmmakers and industry professionals have less time to adapt to the new tools and the possibilities they present. While the technology reshapes the film industry, Hollywood has garnered significant media attention—especially highlighted during events like the well-publicized 2023 Writers Guild of America (WGA) Strike. The labor protest underscored a variety of concerns about how AI is poised to transform creative industries, from scriptwriting to pre-visualization, and hinted at deeper implications for creative authorship and intellectual property within mainstream filmmaking.

However, the focus on the response of Hollywood to AI risks overlooks the more gradual, yet profound, impact AI has already had on video art and artists. Video artists, in particular, have historically embraced technological experimentation, adopting emerging tools from algorithms to Generative Adversarial Networks (GANs) to explore new aesthetic possibilities and challenge traditional modes of representation. This inclination towards boundary-pushing is not new—video artists have long been early adopters, integrating nascent technologies to create content that traditional cinema might deem too experimental or commercially unviable.

Yet, as these AI tools decentralize video production further, they are bringing these innovations to an ever-expanding range of creators beyond the gallery and festival circuit, extending their influence to areas of popular culture, art, and everyday visual media.

Advertising and marketing have become prime areas for AI-driven disruption, harnessing the capacity of the technology for generating eye-catching, short-form, and conceptually simple content. Commercials, by their very nature, are highly adaptable to such tools: they are brief, often rely on fantastical or exaggerated imagery, and focus on delivering clear and memorable messages to audiences. In fact, many of the creatives interviewed for this book, although they may aspire to craft cinematic narratives or experimental video art, work concurrently within marketing and advertising. This dual engagement reflects a broader trend: the broad, transformative impact AI is having on visual culture extends beyond the film industry, permeating art, cinema, popular culture, and advertising alike.

Thus, the influence of the technology in visual media is vast and rapidly expanding, shaping not only how content is created but also how it resonates with audiences across various platforms. This book considers these multifaceted roles played within art, cinema, and advertising, revealing a shift toward more dynamic, responsive, and creatively driven visual culture. Through interviews, case studies, and analyses, readers will discover how AI is enabling new levels of interactivity, creativity, and accessibility in visual storytelling. At the same time, these developments bring forward complex questions about authorship, originality, and ethical considerations, as machine-generated content challenges traditional boundaries in both creative and commercial contexts.

Ultimately, this book underscores the potential of AI not only as a tool for content creation but as a mirror reflecting the collective creativity, biases, and aspirations of society. Through an exploration of the intersections of AI with human expression, readers will gain insight into the evolving role of technology as a collaborator, a disruptor, and a catalyst for new artistic possibilities. Whether within the domain of high art or in everyday advertisements, this transformative reach redefines our relationship with visual media, hinting at a future where creativity and machine intelligence work in tandem to shape the narratives of tomorrow.

Acknowledgements

The authors express profound appreciation to the myriad individuals and institutions whose contributions and unwavering support have been pivotal in the genesis and fruition of this tome. At the vanguard of acknowledgment stands Dr. John Porter, whose innovative leadership in aligning the adoption of emergent technologies like AI with educational initiatives has left an indelible mark on this work's evolution.

Acknowledgment is due to Dr. Kathi Vosevich, Dean of the College of Arts and Humanities, for her steadfast encouragement and support of the research culminating in this publication. Vosevich's guidance and confidence in the authors' academic endeavors have been fundamental in sculpting the contours of this volume.

The authors extend their gratitude to the research team responsible for composing this manuscript and acknowledge the critical insights and contributions of various colleagues and co-authors from ancillary projects. James Hutson extends thanks to Professors Ben Fulcher, Jeremiah Ratican, Joe Weber, Ben Scholle, and Dr. Trent Olsen. As well, Drs. Daniel Plate and Emily Barnes were instrumental in shaping many of the sections and the direction.

The authors also convey their deep appreciation to the administrative leaders at all of the institutions that participated in the study for their relentless support and dedication to creating an environment conducive to academic excellence and scholarly exploration. The leadership, encompassing Presidents, Provosts, and Deans at these institutions, has been instrumental in providing the necessary guidance and resources for this research and writing project.

Furthermore, those that provided insight into their experiences as leaders in industry and education, we are indebted and thank you for your invaluable time and efforts in helping others navigate this unpredictable time. Finally, the authors are thankful for the insights and constructive feedback from colleagues and fellow researchers throughout this project. Their contributions have significantly enriched the work and helped to refine the ideas encapsulated within this volume.

Contents

Chapter 1

Introduction to AI in Video Production

This chapter explores the potential of generative AI (GAI) tools in video production, starting with an overview of current controversies and public discourse surrounding the impact on the industry. It addresses common concerns raised in the media—such as fears of AI replacing human roles and ethical considerations tied to "deepfake" technology—setting the stage for a nuanced exploration of how these tools are reshaping creative fields. The chapter then clarifies the fundamentals of AI-generated video, differentiating generative AI from traditional video production techniques by focusing on its capability to create new visual content autonomously from textual prompts. Providing a brief historical context, key milestones in the evolution of special effects in film will be traced, from early innovations like stop-motion and matte paintings to the introduction of CGI and motion capture, each of which expanded the possibilities of cinematic storytelling. Within this continuum, GAI emerges as the latest revolutionary tool, fundamentally altering the landscape of video production through automation and creative assistance.

1.1 Understanding AI in Video Production Today

The integration of artificial intelligence (AI) into filmmaking and video production is not a recent phenomenon, though its scope has historically been restricted to specific tasks, particularly in video editing and visual effects. AI has long served as a supportive tool, augmenting human creativity and efficiency by automating processes like scene recognition, video analytics, and object detection, enhancing workflows in post-production (Huang et al., 2023; Li, 2024; Sharma et al., 2021). Argaw et al. (2022) have also emphasized the role of the technology in accelerating these tasks, which were once labor-intensive and time-consuming, and in elevating creative possibilities through machine learning-based editing techniques. Such previous scholarship have illustrated the established function of AI as a powerful collaborator within video production, enhancing precision and speed but without fundamentally altering the creative processes driving storytelling.

Recent advancements in generative AI (GAI), however, have redefined the potential applications of AI in film, moving beyond these limited tasks to more autonomous and expansive roles that are transforming the landscape of video creation and distribution. In fact, the technology is increasingly embedded in filmmaking, extending from pre-production through to post-production in ways that streamline workflows, enhance creativity, and reduce costs. In pre-production, AI is employed in scriptwriting and concept development, allowing creators to generate storylines, characters, and dialogue suggestions. Tools like ChatGPT can provide basic frameworks for dialogue or plotlines, while platforms such as ScriptBook analyze scripts for emotional beats and audience appeal, helping filmmakers refine their stories before a single frame is shot (Çelik, 2024). This use of text-based AI in storytelling and planning stages is becoming invaluable, especially for independent filmmakers who might lack a large writing team, as it enables efficient brainstorming and rapid idea development that would otherwise require extensive collaboration (Rahman and Ali, 2024).

In production and post-production, video-based tools are transforming traditional editing and special effects. Even before these latest tools, AI-driven software, such as Adobe's Sensei, was automating tasks like color grading, scene matching, and even background noise reduction, dramatically cutting down the time spent on intricate editing processes (Mugubi and Manono, 2024). Similarly, such software is also used extensively in visual effects (VFX), where machine learning (ML) models enable filmmakers to produce complex CGI sequences and animations more efficiently. GAI tools like Runway Gen-3 and Pictory allow for real-time video enhancements, transforming text prompts into cinematic visuals, which can then be further refined and edited using AI-powered editing software like Descript (Reddy et al., 2024). Additionally, in marketing and audience analytics, AI platforms analyze viewer preferences and demographic data to tailor trailers, advertisements, and even movie endings for specific audiences (Nixon et al., 2024).

On the other hand, while these GAI tools are powerful, they are still in the early stages of development, akin to the evolving nature of large language models (LLMs) in terms of complexity, power, and potential (Tao et al., 2024). Currently, these tools offer capabilities previously limited to high-budget productions, such as creating panning shots, automated transitions, and realistic special effects, all from basic prompts (Hasan et al., 2024). This opens new doors for creators to experiment with complex visuals without requiring extensive technical expertise. However, significant limitations remain: fine control over the nuanced details of scenes, including precise camera movements, actor facial expressions, and specific environmental effects, often remains beyond the reach of these tools (Gerner, 2024). Additionally, many productions still require live actors and human involvement to ensure authenticity and emotional depth, as AI-generated avatars and performances have yet to fully capture the complexity of human expression (Lind, 2024). Despite these constraints, the field is progressing rapidly, with new tools and updates emerging almost daily, bringing incremental improvements that promise to close the gap between human-directed and AI-generated content.

At the same time, the increasing use of such tools has ignited significant debate in both media and industry circles. Much of the discussion centers around concerns that AI may replace traditional human roles, fundamentally altering the creative landscape (Dwivedi et al., 2021). High-profile events, like the Writers Guild of America (WGA) strike of 2023, have highlighted fears within the entertainment industry, with writers and other professionals calling attention to the potential for generative AI tools to disrupt established workflows (Barnes and Koblin, 2023). This discourse underscores concerns over job security, authorship, and creative integrity, as these technologies become more accessible and powerful (Wach et al., 2023).

A particularly controversial aspect of AI in video production is deepfake technology. Deepfakes, which use advanced machine learning (ML) techniques to convincingly simulate human likenesses, have raised complex ethical questions (Masood et al., 2023). On one hand, deepfake technology is being used to creatively extend the capabilities of actors, as seen in projects like David Beckham's multilingual campaign for "Malaria Must Die, So Millions Can Live" (Davies, 2019) and Val Kilmer's post-cancer voice recreation (Etienne, 2021). These uses demonstrate the ability of the technology to enhance media content and increase accessibility by breaking down linguistic and physical limitations.

However, deepfakes also carry risks, such as their potential for misinformation and misuse, including non-consensual applications and reputational harm. These risks are particularly concerning for their impact on privacy and the authenticity of media, prompting some countries to implement restrictive deepfake legislation (Shoaib et al., 2023). A prominent recent example of deepfake use in politics occurred when a manipulated video surfaced of Ukrainian President Volodymyr Zelensky (Fig. 1.1) allegedly calling for his soldiers to surrender to Russia. This deepfake, which appeared on the hacked network Ukrayina 24 and circulated widely on social media, aimed to undermine Ukrainian morale by spreading false information during a tense period of the conflict. The poor quality of the video and distorted visuals made it recognizable as a fake, yet it still caused significant alarm before platforms removed it due to its misleading nature (Allyn, 2022).

Artists like the Dor Brothers are seizing on the current lack of clear regulation around deepfake production, using it as a space for political satire and provocative commentary. Their 2024 video *The Hustle* (https://www.youtube.com/watch?v=aIxqms8KSkA) (Fig. 1.2), depicting famous public figures, including political leaders, robbing a grocery store only to be theatrically arrested, illustrates how AI-generated content can shift beyond simple transactional or experimental pieces to develop intricate narratives and layered critiques. Unlike many deepfake applications aimed at deception, their work leverages humor and absurdity to challenge viewers' perceptions, navigating social and political issues through a comedic lens. While their satirical works may seem exaggerated or whimsical, they underscore serious questions about the ethical implications of deepfake technology. As Yonatan Dor explains, their content serves as a cautionary exploration of "what could happen right now if it's in the wrong hands" (Hutson, 2024).

In recent years, the rise of AI-assisted tools, like those used to create deepfakes, has accelerated the shift toward automating complex production tasks that were once

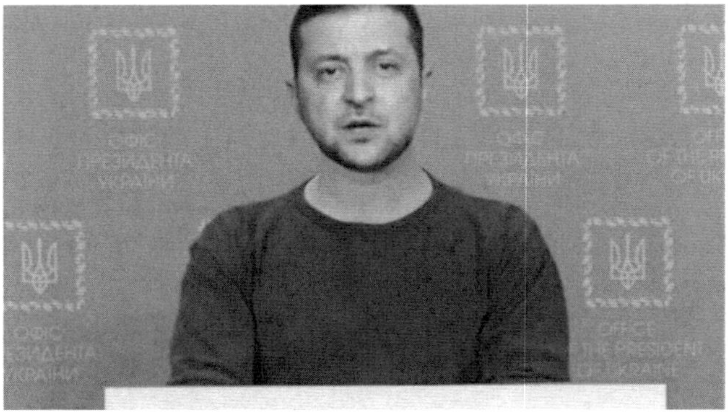

Figure 1.1 A still from the deepfake video of President Zelensky, 2022 (Creative Commons License Zero).

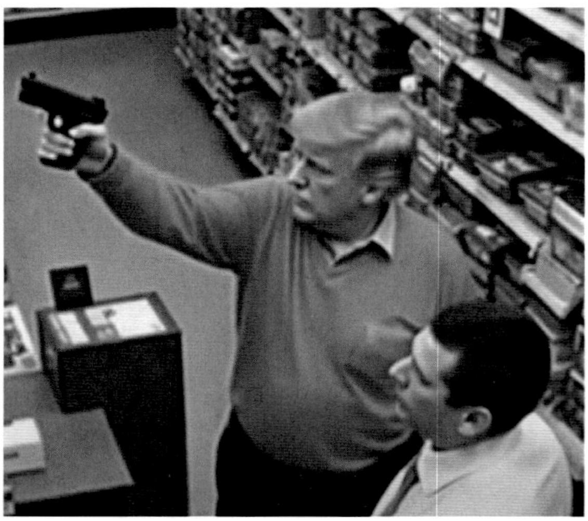

Figure 1.2 The Dor Brothers, *The Hustle*, 2024 (Creative Commons License Zero).

performed by teams of specialists, effectively challenging the traditional roles in video production and filmmaking (Ramaul et al., 2024). AI, an umbrella term, encompasses a range of subfields, each contributing uniquely to the evolving landscape of creative production. Machine learning (ML), natural language processing (NLP), and computer vision (CV) represent three core areas within AI, each with distinct functionalities and applications that converge in the realm of video production. ML, for instance, drives pattern recognition and decision-making processes, making it integral to video editing software that automatically adjusts color grading, audio synchronization, or even editing sequences based on learned aesthetic patterns. NLP, meanwhile, allows AI systems to understand and generate human language, enhancing scriptwriting or enabling systems to interpret user prompts for video generation. CV interprets visual information from images and videos, empowering tools like automatic object tracking, scene segmentation, and facial recognition, which are vital for real-time effects and complex visual edits. Together, these capabilities form the backbone of modern AI in video production, facilitating workflows that allow for greater speed, precision, and creativity (Fteiha et al., 2024).

GAI, a particularly transformative branch within this broader AI ecosystem, refers specifically to systems capable of creating entirely new content, rather than merely analyzing or modifying existing data. GAI leverages advancements in ML, including sophisticated models such as generative adversarial networks (GANs) and diffusion models, to produce original images, audio, and video content. GANs are one of the foundational architectures used in GAI for creating visual content. They consist of two neural networks, the generator and the discriminator, that work in tandem to produce increasingly realistic outputs (Bengesi et al., 2024). The generator's role is to create an image or frame of video from random noise, essentially trying to "fool" the discriminator into believing it is authentic. The discriminator, conversely, attempts to distinguish between real and artificially created content. Through this adversarial process, the generator improves with each iteration, refining the visual quality and realism of the images or video it produces. As a result, GANs are particularly suited to generating complex visual scenes and character animations that look convincingly lifelike (Purwanto et al., 2024).

Diffusion models, another recent and powerful technique, take a different approach by progressively transforming random noise into coherent images or video frames over a series of iterative steps. Diffusion models work by "denoising" images: starting with random noise, the model gradually refines the image according to patterns learned from large datasets, ultimately resulting in a clear and high-quality output. These models excel at capturing intricate details, such as lighting, texture, and motion, which are essential for realistic video production. Diffusion models have become popular in AI video generation because they can handle complex tasks like generating smooth transitions between frames, maintaining spatial and temporal coherence, and preserving scene continuity—challenges that are crucial for believable video content (Po et al., 2024).

The evolution from text-based AI generation to image and, more recently, video generators marks a major shift in creative capabilities across various media over the past three years. Initially, the introduction of text-to-image generators like Stable Diffusion, DALL-E 3, and MidJourney represented a breakthrough in how we conceive of and create visual content. Each of these platforms leverages vast datasets containing millions of images paired with descriptive captions, enabling the models to recognize and replicate correlations between language and visual elements. This foundational process, called "training," equips these models to interpret complex prompts and produce images that closely align with user stylistic and thematic directions (Zhuang and Tang, 2021).

Stable Diffusion (Fig. 1.3), for instance, is renowned for using latent diffusion models to generate intricate, stylized images that often embody a degree of realism paired with artistic flair. It operates through a process that transforms random noise into clear, defined images by iteratively refining details based on learned visual patterns. DALL-E 3, meanwhile, pushes the boundaries of creativity by producing surreal and fantastical imagery that often blurs the line between realism and imagination.

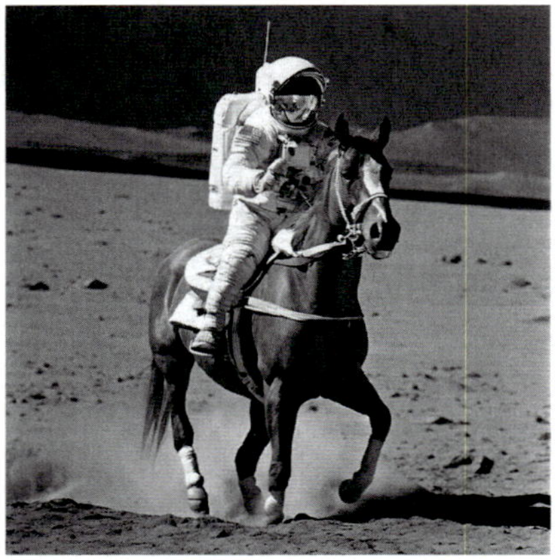

Figure 1.3 VulcanSphere, *Astronaut Riding a Horse*, 2023. Stable Diffusion (Creative Commons License Zero).

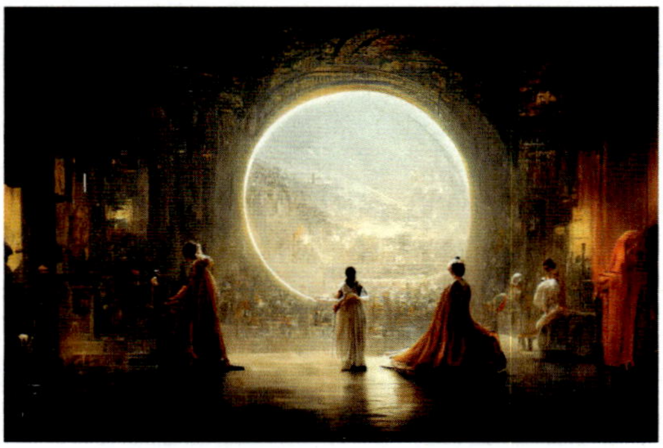

Figure 1.4 Jason M. Allen, *Théâtre D'opéra Spatial*, 2022. Midjourney (Creative Commons License Zero).

This model enables users to craft unique and often unexpected artistic visions by inputting descriptive text, resulting in imaginative outputs that demonstrate the potential to contribute to abstract and conceptual art. MidJourney (Fig. 1.4) further refines these capabilities, generating painterly, cinematic visuals that cater to creators seeking high-quality, artistic aesthetics. With its emphasis on deep tonal variation and immersive visuals, MidJourney appeals particularly to designers, filmmakers, and artists interested in a more traditional, visually rich style (Lee et al., 2023).

The leap from image generators to video generators involves additional layers of technical sophistication, primarily due to the challenges of ensuring temporal coherence and smooth transitions across frames (Liu et al., 2024). Leading platforms like Runway Gen-3, Pictory, and Pika extend the principles of text-to-image generation into video, allowing users to create dynamic sequences based on textual descriptions. Runway, for instance, enables users to create short video clips (https://www.youtube.com/watch?v=HP76MCYkIfU) directly from text prompts, utilizing a similar model structure to text-to-image tools but incorporating temporal models to maintain consistency across frames. Pictory takes a unique approach by converting long-form content, such as blog posts, into visually engaging videos. It analyzes the text for key points, then pairs these elements with corresponding visuals, voiceovers, and animations, effectively transforming static text into narrative-driven video. Pika (https://www.youtube.com/watch?v=NX3oOpO-NP4) differentiates itself with options like 3D animation and cinematic sequences, catering to users aiming for a more filmic or stylized output (Guo et al., 2024).

Both text-to-image and text-to-video generators broaden access to content creation, reducing the need for specialized skills in visual arts or videography. While text-to-image generators focus on capturing single frames that encapsulate visual detail and artistic style, video generators tackle additional complexities: ensuring continuity of motion, consistency in visual details across frames, and coherence in narrative flow. For instance, video models must account for factors like movement, lighting changes, and object interactions over time—features that image models do not need to address. Nevertheless, these tools share a common mission of enabling non-experts to produce high-quality, professional-grade content, opening creative possibilities for individuals with minimal technical knowledge (Wong and Williams, 2024). These platforms illustrate the vast potential to lower the barriers of entry in video production. Independent creators, who may lack access to a traditional resources of studios, can now leverage AI-generated visuals to develop high-production-value content without the need for large teams or specialized skills in CGI or animation (Jin et al., 2024).

This shift has significant implications for the future of video production. The accessibility of GAI tools makes the creative process more accessible, allowing a broader range of voices to participate in the creation of visual media (Torun, 2025). However, as tasks are automated that were traditionally handled by specialists—such as set design, visual effects, and even acting through AI-generated avatars—there is also a fundamental restructuring of labor within the industry. The capacity to perform roles once exclusive to specialized departments suggests a future where creative teams might be smaller, production cycles faster, and budgets lower (Huang et al., 2023). While this accessibility opens up opportunities for diverse storytelling, it also invites discourse on the value of skilled labor and the evolving nature of artistry in a context where machine-generated content becomes increasingly indistinguishable from human-made work. This evolution in AI-driven production methodologies marks a pivotal transition in the visual arts, underscoring the importance of understanding both the possibilities and ethical considerations associated with the role of these tools in shaping the future of video production.

1.2 Reactions to AI Filmmaking: Between Innovation and Controversy

With the announcements of the recent new abilities of these video platforms has come a complex mix of excitement and skepticism, mirroring broader societal debates over the potential and risks of generative technology (Beltran et al., 2024; Wach et al., 2023). The film industry has long served as a testing ground for technological innovations, from CGI to motion capture. However, the introduction of GAI tools, such as OpenAI's Sora, first previewed in February 2024, and Runway's

Gen-1, released in February 2023 with the current Gen-3 model as the premier video generation tool, signals a particularly disruptive wave for the industry (Metz, 2024). With tools that can produce near-cinematic quality content from textual, image, or video-based prompts or conversion, this generative technology stands to lower barriers for entry and streamline pre- and post-production, which has drawn interest from both established studios and indie filmmakers (Rojas and Martínez-Cano, 2024). This initial excitement is underscored by a wave of funding initiatives and partnerships with AI companies, such as Stability AI, OpenAI and Meta, aiming to capitalize on the potential to automate and even enhance the creative process. However, the rapid pace and scale of this technological shift have also raised deep concerns about job displacement, originality, and the evolving role of human artistry within the storied framework of the global film industry (Kulesz, 2024).

This mix of optimism and anxiety is not limited to individual creators but extends into the very structure of the production pipeline of Hollywood. As noted, the Writers Guild of America (WGA) strike of 2023 was, in part, fueled by worries about the growing role of these tools in content generation, as writers feared AI could potentially replace or dilute human storytelling (Clark, 2024). The concerns of the WGA echo those of other guilds and professionals who see AI as a double-edged sword: while it offers unprecedented creative possibilities, it also threatens established industry norms and livelihoods (Kyriakoudes, 2024). With capabilities of generating scripts, assisting in editing, and even designing complex visual effects, industry guilds have called for tighter controls and regulations around AI usage to safeguard artistic integrity and fair compensation (Metz, 2024).

Many industry veterans, such as acclaimed filmmaker Christopher Nolan, have voiced apprehensions about the impact the new tools will have on creative integrity, expressing concerns that generative tools may prioritize efficiency over the nuanced, human-driven storytelling that defines traditional cinema. Nolan, known for his insistence on practical effects and film-based production techniques, argues that an over-reliance on AI could lead to a dilution of cinema's core artistic elements, making films feel manufactured rather than crafted. In the same interview, Nolan noted that these dangers have been apparent for years but have only recently been reported on once chatbots threatened journalists' jobs (Sharf, 2023). Likewise, celebrated director Martin Scorsese has spoken against the overuse of technology, cautioning that AI-driven productions risk becoming formulaic as studios lean toward content that can be quickly generated rather than thoughtfully developed (Shafer, 2024).

This resistance is rooted in past industry shifts, where the rise of accessible CGI often led to an overuse of digital effects, sometimes at the cost of traditional craftsmanship. Several films have been criticized for such overuse, including *Star Wars: Episode I - The Phantom Menace* (1999) faced backlash for its extensive CGI use, especially in characters like Jar Jar Binks and large-scale battles, which some felt overshadowed the practical effects charm of the original trilogy (Fleming and Knee, 2020). *The Hobbit trilogy* (2012–2014) similarly leaned heavily on digital effects for characters and environments, diverging from the practical approach that made *The Lord of the Rings* so beloved, and leaving fans feeling a lack of the grounded realism of the earlier films (Michelle et al., 2017). *I Am Legend* (2007) also saw criticism for its CGI-rendered creatures, which many argued looked unnatural, impacting the immersive nature of the film. The reboot of *The Mummy* (2017) and *Green Lantern* (2011) similarly suffered from CGI-heavy approaches; the digital effects in *The Mummy* diluted the suspense and horror, while *Green Lantern's* CGI suit and environments appeared artificial, detracting from the story (Campbell and Campbell, 2016; Davis, 2018; Essien et al., 2024). Together, these films highlight the risks of overreliance on digital effects, which can sometimes disrupt the immersive quality that practical effects contribute to storytelling.

Likewise, detractors of early generative images and video tools often cited the limitations these technologies displayed, particularly in accurately depicting human features, as evidence of their inadequacy for sophisticated storytelling. Early models released in 2022, for instance, frequently generated distorted or exaggerated facial expressions, and had difficulty with realistic human details like hands, often producing unsettling extra fingers (Fig. 1.5) or warped proportions and faces as in the DALLE Mini example of *Checkers with Chaning Tatum* (2022) (Fig. 1.6). As the technology attempts to mimic forms it cannot fully understand, it inadvertently produces results that feel both familiar and alien (Cassemere-Stanfield, 2023). These visible flaws have continued to circulate on social media as reminders of AI's early challenges, reinforcing skepticism around its viability for mainstream media. Videos like *Will Smith Eating Spaghetti* (2023) (Fig. 1.7) (https://youtu.be/XQr4Xklqzw8) gained viral attention, illustrating the eerie, morphing effect that results when such flawed images are sequenced, producing visuals with an unsettling, fever-dream quality (Grba, 2022). The exaggerated distortions in these early creations quickly became an identifiable aesthetic, sparking further debate about the place of this technology in serious filmmaking.

In response to overreliance on VFX and the unreliability of early video generators, many filmmakers call for the importance of human drama, personal expression and storytelling. Director Quentin Tarantino has consistently highlighted the necessity of infusing films with personal experiences and emotional depth, asserting that his movies are "painfully personal" and that it's his job to make them so, even if only he or those who know him recognize it (Medrut, 2020). This perspective resonates with broader industry apprehensions reminiscent of the CGI era, where an overreliance on digital effects sometimes overshadowed traditional craftsmanship. Similarly, filmmaker James Cameron has expressed concerns about the rapid advancement of artificial intelligence, noting that it is "getting hard to write science fiction" in a world where AI developments are swiftly becoming reality (Balevic, 2024). Such sentiments reflect a broader unease within the film industry regarding the potential for AI to overshadow the human elements that define cinematic art (Green, 2023).

In parallel, major studios have begun to form partnerships with AI-focused companies, seeking ways to integrate these platforms as tools to complement, rather than replace, human creativity. Lionsgate, for instance, has partnered with Runway AI to explore generative tools that support filmmakers in pre- and post-production, aiming to reduce costs while expanding creative possibilities (Lee and Masunaga, 2024). Through the "Hundred Film Fund," Runway has initiated a grant program offering filmmakers up to $1 million to produce AI-enhanced films, highlighting a

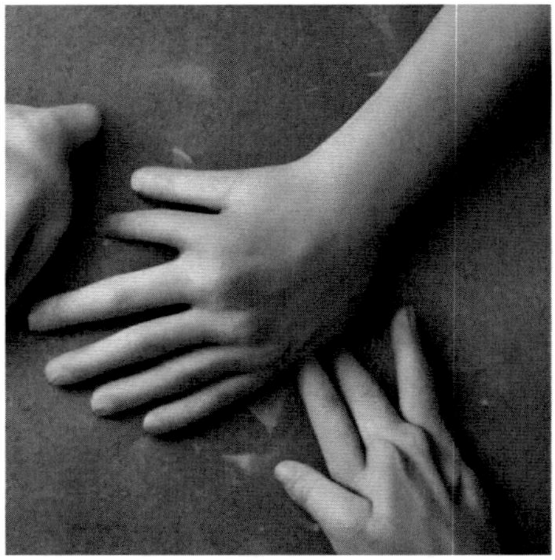

Figure 1.5 Human Hands, Stable Diffusion, 2022 (Creative Commons License Zero).

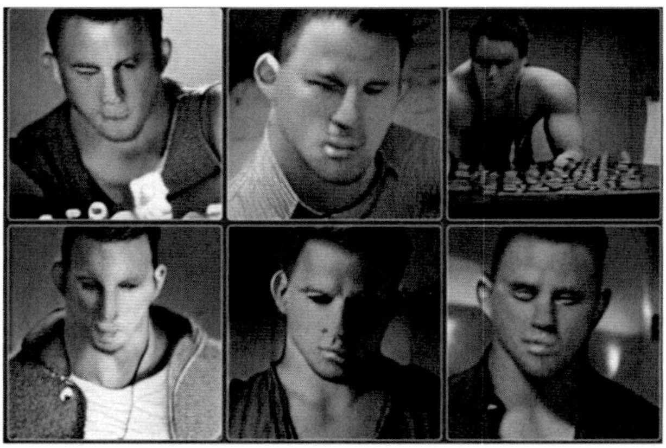

Figure 1.6 Checkers with Channing Tatum, DALLE Mini, 2022 (Creative Commons License Zero).

belief in AI's potential to serve as an empowering addition to traditional filmmaking rather than a substitute (Metz, 2024). While Lionsgate asserts that this technology—built on models trained with their entire film catalog—will be restricted to pre-visualization for greenlighting projects, the potential to leverage the same content for generating scenes during production and post-production is an increasingly foreseeable progression. This collaboration underscores the industry view of these platforms and models as a transformative force that, while collaborative, edges closer to fundamentally reshaping content creation workflows.

At the same time, there have been the entry of high-profile industry figures like James Cameron into the AI sector. Cameron, known for pushing technological boundaries with films like *The Terminator* (1984), *The Abyss* (1989), and *Avatar* (2009), has joined the board of Stability AI, a leading generative AI company known for their image-generation model Stable Diffusion. Despite his earlier lament

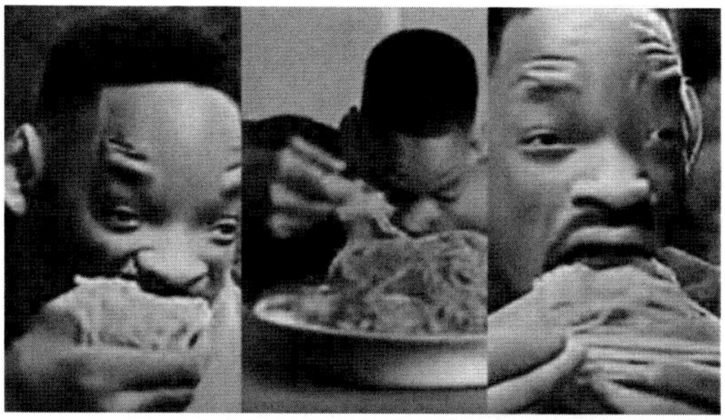

Figure 1.7 Will Smith Eating Spaghetti, ModelScope, 2023 (Creative Commons License Zero).

over how the technology is changing the genre of science fiction, the director has always adopted the latest technologies, often developing new ones in pursuit of his vision. His involvement highlights a growing acceptance within Hollywood's upper echelons, with Cameron advocating for a future where the tools augment creative vision without fully replacing human artistry (Weprin, 2024). This alliance between legacy filmmakers and AI companies is a sign of the normalization of the technology within the film industry; however, it also intensifies the debate over who controls the narrative in an AI-influenced creative landscape.

While these partnerships suggest a hopeful synergy between established filmmakers and AI, the embrace of the technology remains contentious, with ethical and practical issues persistently shadowing its integration (Thomas, 2024). For many filmmakers and creatives, a central concern is whether technology can genuinely replicate the nuanced human touch essential for storytelling—particularly in capturing the subtleties of character development, emotional depth, and thematic resonance that give films lasting impact (Meng et al., 2023). Nonetheless, these generative media solutions have already made inroads into the creative processes of feature films, from CGI enhancements to digital de-aging and voice synthesis, solidifying its place as a pivotal tool in film production and post-production (Murodillayev, 2024). Far from being a future consideration, AI is an active, evolving force within the industry, transforming everything from special effects to audience targeting and marketing. However, the story of the influence of these tools extends beyond film production: it has been quietly reshaping advertising, television, and digital media, establishing itself as a versatile force across the broader landscape of visual culture.

One significant area where GAI has made a substantial impact is advertising and marketing, a field ideally suited for its capabilities in generating visuals, text, and even audio that can be precisely tailored to specific consumer demographics, preferences, and behaviors (Sharakhina et al., 2023). The evolution in marketing parallels the trajectory in video art and filmmaking, where simpler, unimodal tools such as ChatGPT and DALL-E 2 were initially integrated to handle text and static image generation, paving the way for the more sophisticated video components seen today. This integration in advertising is ushering in an era of heightened personalization and efficiency, enabling brands to produce highly individualized content that resonates with consumers on a personal level. By leveraging these advanced systems, advertisers are able to go beyond traditional, static ads, crafting customized, interactive experiences that capture consumer attention and foster deeper engagement (Gao and Liu, 2023). This shift toward AI-enhanced advertising underscores how the technology is simultaneously expanding creative possibilities and streamlining production processes, empowering brands to generate high-quality content rapidly and cost-effectively. Recent examples from major brands highlight the potential use cases of the technology, as it redefines what is possible within the marketing sector.

One early example of the impact in advertising came in 2021 with Cadbury, who launched the *Not Just A Cadbury Ad* campaign (https://www.youtube.com/watch?v=5WECsbqAQSk), utilizing artificial intelligence to support local businesses during the Diwali festival. The campaign featured Bollywood star Shah Rukh Khan, whose likeness was digitally recreated using AI technology to promote small stores affected by the COVID-19 pandemic. By employing machine learning, Cadbury generated personalized advertisements where Khan endorsed local retailers by name, encouraging consumers to shop locally. This innovative approach allowed small businesses to create their own versions of the ad, integrating their store names into the AI-generated content, thereby enhancing community engagement and support. The campaign's success was evident in its widespread reach and impact. Over 130,000 ads were created using the *Shah Rukh Khan My Ad* tool, resulting in more than 30 million views and a 35% growth in business for participating retailers (Salam et al., 2024). This initiative not only showcased the versatility of earlier tools in creating localized and culturally relevant content but also demonstrated its potential to foster community support and drive economic growth for small businesses.

In the consumer goods sector, Heinz has tapped into this creative potential through its partnership with OpenAI's DALL-E 2 for a July 2022 ad campaign (https://www.youtube.com/watch?v=LFmpVy6eGXs). This initiative showcased AI-generated artwork of Heinz ketchup, reinforcing the brand's identity while demonstrating its commitment to innovation. To showcase how well-known the brand is for ketchup, the commercial relays, entirely in text: "Artificial intelligence can now generate images from text. 'Cow Bus.' 'Fry Castel.' 'Cheese Planet.' So we gave them a simple prompt. 'KETCHUP'." The resulting imagery, which is notably before art generators could accurately produce text, produces image after image of the Heinz brand ketchup bottle with illegible text. The visually engaging designs generated with other prompts such as "KETCHUP RENAISSANCE" provided a fresh take on the brand's classic imagery, appealing to a digitally savvy audience while maintaining brand consistency (Matthews et al., 2023). The integration of such content by Heinz demonstrated that the brand was forward thinking, but that the tolls could also streamline the creative process to enhance brand storytelling, setting a precedent for future AI-driven campaigns.

Another such campaign took the tools a step further in actively engage audiences in the creative advertising process. Coca-Cola's *Create Real Magic* campaign (2023) (https://youtu.be/FPGDMj1QUBE?list=TLGGkEhI48FTuZ4wMTExMjAyNA), which combined DALL-E 2 and ChatGPT to create an interactive experience for consumers, invited customers to use generative writing and imaging tools to create their own artwork in collaboration with the brand, blending user-generated content with GAI technology. Through their own interface on their brand website, Coca-Cola integrated the two OpenAI tools with branded examples of themes users could modify from Santa and polar bears to astronauts. Once an example was selected, users could remove background imagery and replace, include new motifs and change the locations. This approach not only encouraged active audience participation but also highlighted the possibilities of co-creation, bridging the gap between consumer creativity and corporate branding (Wright, 2023). Such a campaign underscores the potential to push the boundaries of traditional advertising, fostering a collaborative environment that elevates consumer engagement to new levels.

In the telecommunications sector, Mint Mobile employed ChatGPT in January of 2023 to craft a script for a commercial featuring Ryan Reynolds. In this advertisement, Reynolds explained that he had instructed ChatGPT to write a Mint Mobile commercial in his voice, incorporating a joke, a curse word, and a reminder about the ongoing holiday promotion by the company. He then read the generated script on camera, describing the result as "eerie" and "mildly terrifying" due to its uncanny resemblance to his own style (Beer, 2023). The approach, like that with Heinz and Coca-Cola, not only showcased creative potential but also added a layer of authenticity and humor that resonated with viewers. Moreover, as with other brands, the transparent integration of AI into the ad's creation highlighted the nuanced relationship between AI and human creativity in content production without violating consumer faith in the products (McAlister et al., 2024).

While these examples demonstrate the viability of generative text and image integration into ads, others showcase broader immersive appeal. For instance, Virgin Voyages' *Jen AI* campaign (https://www.youtube.com/watch?v=H-B5mB53bgc) of 2023 featured an AI-generated avatar of Jennifer Lopez, inviting consumers to book cruises and interact with a digital representation of the star. The success of the campaign required partnering with creative agencies VMLY&R and Deeplocal, who were able to use custom avatars to deliver a uniquely personalized experience that allowed audiences to engage directly with a celebrity figure in a way that would be difficult to achieve through traditional means (Campbell, 2023). For instance, customers could send an invitation to a friend to join them on a flight and the avatar of JLo would use the names of both individuals. Instead of having the actress read thousands of potential names with varying pronunciations, the use of voice synthesis combined with lip syncing tools had the virtual version of her seamlessly communicate directly to customers. This is one such campaign that exemplifies how these tools can forge personal connections, offering brands a powerful tool to create memorable and interactive customer experiences (Thomas, 2024).

In 2024, Toys 'R' Us ventured into the realm of AI-generated advertising by releasing a commercial created using OpenAI's Sora, a text-to-video tool (https://youtu.be/F_WfIzYGlg4). This minute-long film narrates the origin story of the company's founder, Charles Lazarus, and its mascot, Geoffrey the Giraffe, entirely through AI-generated visuals. The creative agency Native Foreign collaborated with Toys 'R' Us to produce this advertisement, utilizing Sora to generate realistic scenes and characters based on textual prompts (Sager, 2024). While the video platform handled the bulk of the visual creation, human intervention was necessary for corrective visual effects and the addition of an original music score composed by Aaron Marsh of the indie-rock band Copeland. The commercial premiered at the 2024 Cannes Lions Festival, signaling a significant shift in advertising practices and setting a precedent for other brands to explore AI-driven content creation (Edwards, 2024).

These cases illustrate the expansive potential of generative technology in transforming contemporary advertising, underscoring its ability to reshape marketing approaches and elevate consumer engagement. The empowerment of brands to produce personalized, high-caliber content that aligns with specific audience preferences represents a notable pivot towards enhanced creativity and individualized communication. The trajectory not only foregrounds the efficiencies introduced by generative AI but also signals a broader reconfiguration of brand-audience dynamics across diverse media platforms and cultural settings. In essence, target audiences organically can engage with brands now by creating their own content and sharing on social media to reach their networks and so on. Thus, as technology advances, its impact on advertising is poised to deepen, creating avenues for brands to design innovative, resonant campaigns with broad, cross-cultural appeal (Sahin and Soylemez, 2024).

The shift in advertising foreshadows a parallel evolution within the realm of film, starting with short-form film, where AI is similarly redefining creative processes and expanding access to visual storytelling. Here, too, generative technology facilitates a popularization of content production, allowing independent creators and established studios alike to leverage AI for both artistic experimentation and narrative depth. Unlike traditional filmmaking methods, these new video platforms enable the production of visually compelling narratives with reduced crew requirements and minimal setup time. The shift in short-form film production can be understood as moving from traditional methods to generative AI-driven approaches is emblematic of a broader transformation in visual storytelling, mirroring the trajectory seen in advertising. Initially, creators adopted a blended approach, integrating both conventional techniques and generative ones to expand creative possibilities.

The hybrid model allowed filmmakers to experiment within familiar frameworks, gradually incorporating the capacity for generating complex visuals using emergent platforms and enhancing narrative depth. The sci-fi miniseries produced by DUST *Sigma_001* (2023) (https://www.youtube.com/watch?v=V7ONOfg2Nzs) directed by Quinn Halleck embodies this balanced approach, using generative AI to construct intricate, futuristic worlds while maintaining traditional filmmaking techniques. The early versions of video generators were unpredictable and could not realistically portray human actors, emotions and situations. Therefore, these early examples maintain human-directed cinematography with AI-generated establishing shots, special effects, and cut scenes. In such a way, *Sigma_001* exemplifies how early integrations served as a powerful complement to conventional methods, enriching storytelling without overshadowing the collaborative essence of filmmaking.

Moving beyond hybrid methods, some projects have embraced fully generative platforms, leveraging them to explore the bounds of visual cohesion and stylistic consistency across extended narratives. *The Frost* (2023) (https://www.thefrostpart.

one/), a 12-minute short film was created entirely with DALL-E 2 generated imagery was collaboratively developed by Waymark, a Detroit-based video creation company, and Latent Cinema. The film's narrative unfolds in the icy depths of Antarctica, where a team investigates a mysterious signal, embarking on a journey that challenges their understanding of past and future events. To animate the still images, filmmakers utilized D-ID, a service specializing in adding motion to static images. The technology enabled the creation of lifelike movements, such as characters speaking and walking, thereby enhancing the storytelling and immersing the audience in the film's world. However, this early, bizarre example of AI-generated film demonstrates the challenges of working in this new medium, namely maintaining shot continuity and conveying emotions. The team faced difficulties in capturing complex emotions like fear, terror, sadness, hopelessness, and elation in the characters while ensuring convincing interactions. Additionally, animating still imagery to breathe life into the visuals necessitated innovative approaches, such as using Mixamo 3D characters for movement, particularly during walking scenes (Heaven, 2023).

The 2023 animated short film *Critterz* (https://www.critterz.tv/) exemplifies the evolving role of artificial intelligence in animation production. Directed by Chad Nelson, the film represents the first animated short completely designed by AI-generated visuals using OpenAI's DALL-E, including characters and backgrounds, directly from textual descriptions. The method allowed the production team to bypass traditional design processes, significantly reducing the time and resources typically required for animation. By inputting specific prompts, the creators produced a cohesive and imaginative world, demonstrating the capacity to translate abstract concepts into detailed visual representations. This approach not only accelerates production but also opens new avenues for creative expression, as AI-generated visuals can inspire novel artistic directions. As such, *Critterz* serves as a compelling case study in the integration of AI into the creative industries, highlighting both the opportunities and challenges that accompany this technological advancement.

For filmmakers exploring the intersection of human and AI-driven aesthetics, *Thank You for Not Answering* (2023) (https://vimeo.com/groups/shortfilms/videos/821101511), directed by Paul Trillo, illustrates the collaborative potential of generative AI in filmmaking by integrating Runway Gen-2 to produce animated sequences derived from text and image prompts. Through the capabilities of the platform, Trillo crafted visuals that align with the themes of memory and perception, blending AI-driven surreal imagery with creative direction. The whole of the film is a missed call by a man overlayed on surrealistic imagery of men and women in, on or surrounded by water as a metaphor for the melancholic regrets of memory. The method highlights the collaborative force such tools can represent in the creative process, enabling filmmakers to experiment and iterate on aesthetic ideas with speed and flexibility. At the same time, the tell-tale eerie distortions and blurred movements capture the early aesthetic of AI-generated video.

In the realm of documentary filmmaking, *CHECKPOINT: Creativity in the Age of AI* (2023) (https://vimeo.com/801101673) made by Áron Filkey and Joss Fong for SPACE10 takes these capabilities a step further by using them to produce both narration and visuals, crafting a meta-narrative on the nature of creativity itself.

Through sequences of text overlaid on short relevant clips and narrated throughout, the evolution of creativity charts the development of human creativity to "creative machines." Through the blurring of the boundary between creator and machine, *CHECKPOINT* emphasizes the potential to add depth to documentary storytelling, prompting viewers to question the origins of creativity and the extent of machine influence. This experimental use of AI in documentary not only enriches the genre but also challenges traditional perceptions of authorship and originality in film.

Finally, *HYSTERESIS* (2023) (https://youtu.be/YGtRnSD2amA) by Robert Seidel showcases the avant-garde potential of these tools by using it to remix classical imagery, creating an abstract, immersive experience that defies conventional narrative structures by leaning into the distortions common to the tools. The psychedelic imagery evolves from the non-objective and abstract to the humanoid through the distortive qualities of this generation of models. This award-winning short film exemplifies how AI can disrupt and transform established art forms, reimagining familiar visuals in innovative ways. Through the unpredictable lens of generative AI, *HYSTERESIS* challenges viewers to engage with the visual medium from a new perspective, suggesting that the evolving technology could inspire entirely novel genres and aesthetics in filmmaking. Collectively, these films illustrate how AI-driven video production is diversifying cinematic expression, from animation to documentary and beyond, offering filmmakers a suite of creative tools to explore uncharted artistic landscapes.

At the same time, intelligent video solutions have profoundly permeated not only Hollywood, but the global film industry, particularly within visual effects and character augmentation, reshaping traditional production processes and pushing artistic boundaries. A prominent example lies in Marvel's *Secret Invasion* (2023) (https://www.youtube.com/watch?v=iqnBJHU7si4), where generative techniques were employed to produce visually fluid, morphing imagery within the opening sequence, capitalizing on the distortive qualities seen in videos like *Will Smith Eating Spaghetti*. Developed by Method Studios, these custom systems were instrumental in crafting visuals that captured the transformative essence of the Skrulls, a central theme in the narrative of the series. This application sparked substantial debate within the industry: while some critics voiced concerns over the potential sidelining of conventional artistry in a post-truth era of simulacrum, others contended that such technology enriches creative horizons (Pramanik and Rai, 2023).

Beyond single sequences, automated visual technologies have seamlessly integrated into entire productions, as evidenced in *The Matrix Resurrections* (2021). Here, algorithms enabled the creation of intricate scenes blending live-action footage with computer-generated imagery, achieving a fluidity and coherence that might be challenging to replicate through manual effects alone. This technology allowed for more dynamic interactions between physical and digital environments, demonstrating how advanced models can effectively bridge the gap between reality and simulation (Essien et al., 2024). Similarly, *The Lion King* (2019) remake leveraged generative approaches to create lifelike animal behaviors, enhancing the immersive qualities of the film. By simulating nuanced animal interactions and natural movement, the film exemplifies the potential of this technology to elevate visual storytelling, adding layers of realism beyond what traditional CGI could achieve.

The impact of generative media solutions has also extended into the realm of actor age manipulation, where tools now enable seamless de-aging across a character's narrative timeline. For example, *The Mandalorian* (2019) and *The Book of Boba Fett* (2021) employed Respeecher to digitally recreate the voice of a young Luke Skywalker, enabling a continuity in sound that aligned with earlier portrayals of the character. *The Irishman* (2019) took a similar approach, utilizing AI-driven de-aging processes on Robert De Niro and Al Pacino, bypassing the need for conventional tracking markers on faces of the actors (Loock, 2021). Disney's development of the Face Re-Aging Network (FRAN) further illustrates this evolution, as this tool offers an efficient and accurate means to alter age features, keeping characters consistent in their visual timelines (Farinacci, 2024).

Beyond age manipulation, the ability to digitally recreate actors posthumously has opened new possibilities while generating considerable ethical discourse. The 2016 film *Rogue One: A Star Wars Story* employed advanced digital techniques to recreate the likenesses of deceased actor Peter Cushing as Grand Moff Tarkin and a younger version of Carrie Fisher as Princess Leia. This practice has sparked significant ethical debates regarding the posthumous use of actors' images. In particular, the estate of Peter Cushing initiated legal action against Disney and Lucasfilm, alleging unauthorized use of his likeness. The lawsuit contends that the digital resurrection violated agreements made prior to Cushing's death, highlighting the complex legal and moral considerations surrounding the use of artificial intelligence in film production (Bacon, 2024). Likewise, *Fast & Furious 7* employed similar tools to portray Paul Walker following his untimely passing, merging body doubles with digital augmentation to complete his character's narrative. The upcoming *Finding Jack* has announced plans to digitally recreate James Dean, prompting further dialogue on the ethical dimensions of these technologies in reviving actors. Such instances underscore the duality of these tools: while they enable continuity and pay homage, they also necessitate careful consideration of moral implications.

Together, these applications reveal the expanding role of automated video systems within cinema, showcasing their ability to enrich and transform the visual landscape of film. However, the rise of such technologies also calls for a balanced approach that respects traditional craftsmanship and acknowledges ethical concerns. As the industry continues to embrace these advancements, cinema remains an evolving arena where technological innovation and human artistry converge, suggesting a future where these tools will coexist with—and potentially redefine—the art of filmmaking.

1.3 A Brief History of Special Effects in Film and Video

The tradition of visual effects in film underscores a continuous push for innovation that has evolved dramatically since the first film in 1888, each technological leap redefining how narratives are told on screen. In the earliest days of cinema, filmmakers began with practical effects and stage-like setups, capturing actors against stationary backdrops, reminiscent of theater performances—a straightforward transition from public entertainment of the time (Rosenstone, 2017). However, as demand for immersive storytelling grew, filmmakers began to explore techniques that expanded beyond static recordings. These first filmmakers of the late nineteenth century often emerged from backgrounds rooted in science, engineering, and the visual arts. Figures like Thomas Edison (1847–1931) and the Lumière brothers, Louis (1864–1948) and Auguste (1862–1954), leveraged technological innovations of their time to experiment with moving images. Edison, who held a fascination with motion and mechanical invention, developed the Kinetoscope, a device allowing individuals to view short, moving scenes through a peephole, which became popular in the early 1890s (Dickson and Dickson, 2000).

The Lumière brothers, initially involved in the photographic business of the family, shifted towards motion picture technology after witnessing Edison's Kinetoscope, creating the Cinématographe, a lighter, more versatile camera that could also project films onto a screen, which played a crucial role in public film exhibitions beginning in 1895. Their scientific curiosity and technical expertise in capturing images laid the groundwork for cinema as both an art form and a commercial enterprise (Gardies, 2000). Additionally, early filmmakers like Eadweard Muybridge (1830–1904), a photographer, and Étienne-Jules Marey (1830–1904), a physiologist, contributed to this foundational period by using sequential photography to analyze motion, which prefigured film techniques (Cresswell, 2012). This eclectic mix of scientific, theatrical, and commercial perspectives established a foundation for cinema that embraced both innovation and spectacle, a dynamic that continues today with the latest generation of creatives using generative media coming from industries outside of traditional filmmaking, such as graphic design, photography, video editing, and advertising.

Moreover, as the medium matured in the early 1900s, the technical considerations of the moving picture gave way to another endeavor: special effects. Theatrical performers such as Georges Méliès (1861–1938), a former stage magician, brought a sense of wonder and illusion to early films, using stop-motion and multiple exposures to create surreal effects in movies like *Le Voyage dans la Lune* (*A Trip to the Moon*) (1902) (Fig. 1.8), which bridged theater and film with fantasy elements. The former illusionist applied stage magic principles to develop groundbreaking cinematographic techniques: multiple exposure allowed layered imagery through repeated film exposure, while stop-motion created apparent movement through frame-by-frame manipulation. The theatrical practitioner constructed elaborate sets reminiscent of stage designs, yet enhanced through camera techniques impossible within traditional theater confines. Multiple exposure effects enabled celestial bodies to display human faces, while stop-motion allowed spacecraft to seemingly emerge from cannon barrels and crash-land into lunar surfaces. Hand-painted frames added chromatic vibrancy to selected prints, merging artisanal craftsmanship with mechanical reproduction (Crivelli, 2023).

As special effects advanced, matte painting became another key technique, enabling filmmakers to create expansive settings beyond the physical limitations of studio backdrops. By the 1930s and 40s, matte paintings allowed filmmakers to depict scenes on a grander scale, often blending these artistic creations seamlessly

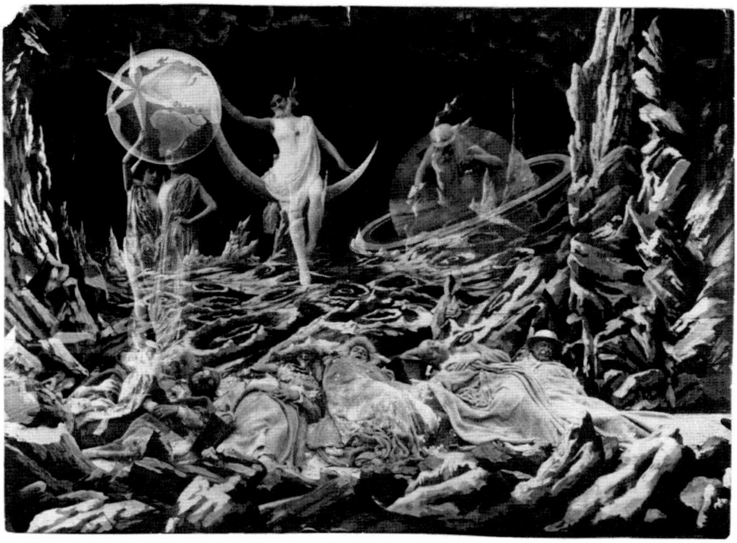

Figure 1.8 Georges Méliès, *A Trip to the Moon*, 1902 (Creative Commons License Zero).

with live-action footage (Mitchell, 2013). Pioneering matte artist Albert Whitlock (1915–1999) revolutionized cinematic landscapes through meticulous attention to architectural and atmospheric detail. Working on Alfred Hitchcock's *The Birds* (1963), Whitlock created haunting vistas of Bodega Bay through glass paintings placed strategically before cameras, establishing spatial depth while maintaining photographic credibility. The technique involved painting portions of scenes on glass while leaving specific areas transparent, allowing live action elements to merge seamlessly with painted environments (Lanza, 2020). Likewise, the contributions of Peter Ellenshaw (1913–2007) to numerous Disney productions exemplified narrative power of the approach. For *Mary Poppins* (1964), Ellenshaw crafted expansive Victorian London roofscapes, enabling characters to dance across chimney tops through precise alignment of painted architectural elements with physical sets. The optical printer, operated by British effects supervisor Eustace Lycett (1914-2006), permitted multiple pass photography—combining live-action footage, painted elements, and animated sequences into unified compositions (Kurop, 2003).

Matte painting reached new heights during *Star Wars* (1977), where American artist Ralph McQuarrie (1929–2012) constructed vast planetary landscapes and space stations. The expansive corridors of the Death Star emerged through strategic combination of minimal physical set pieces with painted extensions. The optical printing innovations of Industrial Light & Magic, supervised by another American artist, John Dykstra (1947–), enabled unprecedented integration of miniature photography with matte elements (Rehak, 2018). These techniques established frameworks later adapted for digital matte painting in films like *Blade Runner* (1982), where the conceptual designs of Syd Mead (1933–2019) materialized through the atmospheric cityscapes of Matthew Yuricich (1923–2012) (Frelik, 2022). Finally,

Raiders of the Lost Ark (1981) demonstrated the capacity of matte painting for period authenticity. American art director Michael Pangrazio (1956–) created ancient temples and vast warehouse interiors through subtle gradations of light and shadow, while optical printing supervisor Bruce Nicholson (1948–) orchestrated seamless transitions between practical locations and painted extensions, which he had just used while working on *Star Wars: The Empire Strikes Back* (1980). The warehouse final shot sequence, revealing endless rows of stored artifacts, exemplified matte painting's ability to expand narrative scope beyond physical production constraints (McNeese, 1989).

The introduction of computer-generated imagery in the 1970s and 1980s marked a seismic shift in film production. Early CGI experimentation manifested through pioneering work by John Whitney Sr. (1917–1995), whose mechanical analog computer generated unprecedented abstract patterns for the opening credits of Hitchcock's *Vertigo* (1958) (Walley, 2022). *Westworld* (1973) marked significant advancement through raster graphics processing—supervisor John Whitney Jr. (1946–2021) utilized digital image processing to simulate android point-of-view sequences, creating pixelated thermal vision effects through frame-by-frame scanning and color manipulation (Venkatasawmy, 2012).

The 1982 science-fiction film *Tron* revolutionized CGI implementation through collaborative efforts between Triple-I, MAGI/Synthavision, Robert Abel & Associates, and Digital Effects. Supervisor Richard Taylor (1944–) orchestrated unprecedented computer-generated sequences, requiring thirty-two minutes of fully digital animation. The light cycle sequence emerged through MAGI's SynthaVision system, which utilized solid geometric shapes called primitives to construct three-dimensional objects (Prince, 2011). Mathematical algorithms guided vehicle movements while specialized rendering software created distinctive illuminated effects.

While these early milestones were pivotal, the innovations of Industrial Light & Magic for *Jurassic Park* (1993) fundamentally altered cinematic possibilities. Visual effects supervisor Dennis Muren (1946–) collaborated with seasoned stop-motion animator Phil Tippett (1951–) to develop revolutionary animation techniques. Traditional go-motion puppetry, including a life-sized T-Rex head and torso, merged with digital elements through custom software development. The T-Rex sequence combined full-scale animatronic elements from Stan Winston Studio with CGI rendered through Softimage 3D software, achieving unprecedented organic movement and textural authenticity (Shay and Duncan, 1993). Later in the decade, *The Matrix* (1999) established new paradigms through visual effects supervisor John Gaeta's (1965–) bullet time innovations. The technique required precise configuration of 120 still cameras and two motion picture cameras arranged in circular arrays. The sequence that was to be parodied *ad nauseum*, captured protagonist Neo seemingly moving so fast that he could dodge bullets. Custom interpolation software developed by Manex Visual Effects generated intermediate frames, creating fluid movement through space while maintaining temporal distortion (Rehak, 2007). Digital compositing through Flame and Inferno software integrated multiple elements seamlessly.

Motion capture (MoCap), which gained prominence in the 1990s and early 2000s, allowed for a new level of realism in character animation by capturing the nuanced movements of actors. Motion capture bridged the gap between animation and live action, impacting the realism of digital characters in blockbusters like *The Lord of the Rings* triology (2001–2003). MoCap introduced new editing and compositing challenges, as animators had to blend captured motion data with CGI characters, creating a fusion of real and digital performances (Bregler, 2007). Special effects *avant-garde* director James Cameron (1954–) advanced MoCap technology through virtual camera systems within the much anticipated *Avatar* (2009). Effects supervisor Joe Letteri (1957–) led Weta Digital toward groundbreaking facial capture technology, as head-mounted cameras recorded minute performer expressions. Data generation reached unprecedented volumes—a single animation frame contained 1.4 gigabytes. Muscle simulation software created realistic character movement while volumetric lighting systems rendered photorealistic interaction between digital characters and environments (Katz and Ellis, 2024). The facial capture system established new standards for performance translation, leading to widespread adoption across major studios such as Industrial Light & Magic and Digital Domain. The volumetric capture stage, a 360-degree array of infrared cameras recording surface deformation and spatial dynamics, became standard protocol for subsequent blockbusters including The *Jungle Book* (2016) and *Ready Player One* (2018), while developments in real-time rendering through game engines like Unreal enabled directors to view completed effects during principal photography (An, 2022).

The latest chapter in this evolution is increasing utilization of various AI technologies for various special effects purposes, which stands to revolutionize the film industry yet again. This leap forward is built upon the foundation laid by CGI but introduces a new level of automation and versatility. Key technological breakthroughs in ML and neural networks have been instrumental in this transition. GANs, for instance, have enabled the creation of photorealistic images and videos from textual descriptions, offering filmmakers tools to visualize scenes and concepts rapidly. For instance, Meta AI developed StyleGAN3 in 2021, enabling unprecedented facial animation control through adaptive discriminator augmentation, while Stability AI launched Stable Diffusion, opening up image generation through latent diffusion models (Karras et al., 2021). Along the same lines, DeepMind advanced motion synthesis through reinforcement learning, creating natural character animations from limited reference data.

Visual effects studios radically transformed production methodologies through artificial intelligence integration. As noted, Industrial Light & Magic developed proprietary deep learning algorithms for facial reconstruction, enabling seamless age manipulation of actor Mark Hamill through neural rendering in *The Mandalorian* (2019–2024). The process analyzed thousands of reference photographs, creating detailed facial maps while preserving micro-expressions and subtle performance nuances. A specialized convolutional neural network processed frame-by-frame adjustments, maintaining temporal consistency across sequences (Yang et al., 2023). Similarly, Weta FX advanced environmental generation through implementation of custom GAN architectures for *The Batman* (2022). The system processed extensive photogrammetry data of Gothic architecture, generating procedural variations of Gotham City environments. ML algorithms analyzed lighting conditions from principal photography, automatically matching atmospheric conditions and maintaining visual continuity. Neural rendering enabled real-time environment modifications during virtual production, facilitating director Matt Reeves' (1966–) creative decisions through immediate visualization (Bratter, 2022).

One of the most pervasive instances of AI use in features saw frame interpolation achievements in *Everything Everywhere All at Once* (2022), which emerged through novel applications of ML models. OpenAI collaborated with A24 to develop adaptive neural networks capable of generating intermediate frames without motion blur artifacts common to traditional techniques. The system analyzed action sequences at molecular levels, preserving clarity in rapid transitions between parallel universes. Directors Daniel Kwan (1988–) and Daniel Scheinert (1987–) utilized the technology to achieve complex temporal manipulations while maintaining production efficiency within independent film budget constraints (Yang et al., 2023). Luma Pictures pioneered AI-driven crowd simulation for urban sequences, implementing deep learning models trained on human locomotion data. The system generated naturalistic background action while reducing computational overhead. Similar advances emerged through Digital Domain's neural network-based fluid simulation system, enabling complex water and particle effects through automated processing of physical dynamics. The next year, Runway ML introduced text-to-video generation capabilities, enabling directors to generate preliminary visualization sequences through natural language prompts. The technology streamlined pre-visualization processes, reducing production costs while expanding creative possibilities. Concurrent developments in neural rendering enabled real-time environment modification during virtual production, as evidenced in The Volume stage system utilized for Ahsoka (2023) (Anderson, 2024).

While major attention has been paid to the inroads and impact AI has been having in film that advances in video art and marketing are being overlooked. For example, contemporary experimental artists propelled technological innovations beyond traditional cinema boundaries. Bill Viola (1951–) transformed video installation practices through StyleGAN integration, creating responsive environments exploring human consciousness at Venice Biennale 2023. Cao Fei (1978–) advanced digital identity examination through custom neural networks, generating evolving avatar performances responding to viewer presence. *Synthetic Realities* at Whitney Museum (2023) established new paradigms through artists manipulating GANs and neural networks to examine authenticity and perception in digital spaces (Giacomin Da Silva, 2023). Major technological advances extended across multiple sectors. Research laboratories including MIT Media Lab and Samsung Research developed specialized neural architectures for generative art applications. Medical imaging benefited from enhanced visualization techniques originally developed for cinema. Architectural firms adopted environmental generation systems for urban planning, while aerospace companies implemented simulation technologies for advanced testing protocols (Hasselgren and Oprea, 2024). WPP and Publicis developed sophisticated demographic analysis systems enabling real-time content modification.

Neural networks processed viewer data streams, automatically adjusting visual elements for maximum engagement. Fortune 500 companies including Nike, Apple, and Microsoft established new communication paradigms through AI-generated campaign assets, fundamentally altering advertising methodologies while raising critical questions about authenticity and artistic ownership in digital media creation (Labrecque et al., 2024).

Looking back over the history of special effects in film and video, there have been many milestones and inflection points for creative industries. However, the most recent emergent technologies fundamentally alter traditional production paradigms, marking an unprecedented empowerment of creative capabilities. Recent developments through 2023–2024 demonstrate radical acceleration in accessibility, as tools previously requiring extensive technical expertise become available to independent creators. Neural networks now accomplish tasks which demanded years of specialized training, enabling rapid ideation and execution of complex visual narratives. The result is the most disruptive period in creative history for all modalities from images, video, music, and more. Given that film brings all of these to bear in storytelling, the industry is being pressured from multiple specializations. To understand the extent to the impact, one merely needs to consider an historical examination of the industry which reveals distinct evolutionary phases in special effects development—from practical effects through CGI to current AI integration. Each phase expanded creative possibilities while maintaining significant barriers to entry through equipment costs and technical requirements. Contemporary AI systems, however, disrupt established patterns, providing sophisticated capabilities through accessible interfaces. Independent creators now generate effects previously requiring substantial studio resources, fundamentally altering production methodologies and creative approaches.

Previous publications examining the phenomenon of AI applications in video production have focused primarily on corporate implementation and technical specifications (Babock and Bali, 2021; Foster, 2022; Sawant, 2024). The present volume diverges through examination of emerging creative applications, documenting experiences of independent artists and creators previously excluded from high-end production processes. Case studies demonstrate unprecedented decentalizing of video creation, analyzing projects completed through minimal technical infrastructure yet achieving professional quality standards. As a matter of fact, those interviewed for the manuscript unanimously agreed that the most disruptive and promising aspect of the technology is allowing one creator to produce a feature length film on their own, controlling the script, photography, virtual actors, music, editing, special effects. We are poised at the edge of a new age of potential for longform storytelling that can bypass traditional channels of film distribution, leading to a wider array of perspectives, voices, and audiences. The volume at hand thus provides a comprehensive analysis of current AI video production capabilities while acknowledging rapid technological evolution. Emphasis remains on practical applications and creative methodologies rather than specific software implementations, ensuring continued relevance as tools advance. Direct creator perspectives illuminate novel approaches to narrative development, visual design, and production workflow optimization.

1.4 Overview of the Manuscript

The next six chapters provide an overview of the current and future impact of the latest generative video tools on visual media, particularly within the realms of film and video art. It systematically examines the technological foundations of AI-generated visuals, situates these advancements within historical and theoretical contexts, and provides practical guidance for creators. Through detailed case studies and analyses, the work explores the evolving relationship between AI and human creativity, offering insights into the future trajectory of AI in visual culture.

Chapter 2 provides an in-depth analysis of the primary platforms facilitating AI-generated visuals, including Stable Diffusion, MidJourney, and Runway. It elucidates the underlying mechanisms of Generative Adversarial Networks (GANs) and diffusion models, highlighting their roles in image and video synthesis. Additionally, the chapter addresses the technical challenges inherent in these technologies and explores the opportunities they present for innovation in visual media.

Chapter 3 traces the evolution of digital art and video experiments, emphasizing the early intersections of cinema and AI in visual storytelling. It examines various theoretical frameworks, including aesthetic movements, to assess AI's influence on contemporary visual narratives. By situating current AI applications within a broader historical and theoretical context, the chapter offers a comprehensive understanding of AI's role in shaping visual culture.

Chapter 4 profiles video artists recognized as pioneers in AI video art. It presents case studies that illustrate AI's function as a collaborative partner in creative processes, and analyzes works that employ satire, surrealism, and social commentary through AI-generated visuals. These examples underscore the diverse applications of AI in contemporary video art.

Chapter 5 serves as a practical manual for creators interested in utilizing AI tools for video production. It offers detailed instructions on generating AI art from text prompts using platforms like MidJourney and DALL·E, complete with sample prompts and step-by-step processes accompanied by screenshots. The chapter also covers techniques for animating still images using platforms such as Emu and Sora, and provides guidance on integrating AI-generated content into traditional video editing software like Adobe Premiere Pro and Cinema 4D for advanced post-production enhancements.

Chapter 6 explores emerging trends in AI-driven visual media, with a focus on advancements like Runway Gen-3. It discusses the potential future intersections of art, technology, and society, and envisions AI's expanding role in broadening creative expression. The chapter aims to provide a forward-looking perspective on the evolving landscape of AI in film and visual culture.

References

Allyn, B. (2022). Deepfake video of Zelenskyy could be 'tip of the iceberg' in info war, experts warn. *NPR.* March 16, 2022: https://www.npr.org/2022/03/16/1087062648/deepfake-video-zelenskyy-experts-war-manipulation-ukraine-russia.

An, D. (2022). Technology-driven virtual production: the advantages and new applications of game engines in the film industry. *Revista Famecos*, 29(1): e43370–e43370.

Argaw, D. M., Heilbron, F. C., Lee, J. Y., Woodson, M. and Kweon, I. S. (2022, October). The anatomy of video editing: A dataset and benchmark suite for ai-assisted video editing. pp. 201–218. *In*: *European Conference on Computer Vision*. Cham: Springer Nature Switzerland.

Babcock, J. and Bali, R. (2021). *Generative AI with Python and TensorFlow 2: Create Images, Text, and Music with VAEs, GANs, LSTMs, Transformer Models*. Packt Publishing Ltd.

Bacon, T. (2024). Lucasfilm sued over *Rogue One's* use of Peter Cushing as Grand Moff Tarkin. Screen Rant. September 11, 2024: https://screenrant.com/star-wars-rogue-one-grand-moff-tarkin-cameo-lawsuit/.

Balevic, K. (2024). James Cameron told Bill Gates that it's 'getting hard to write science fiction' in the world of AI. *Business Insider*. September 18, 2024: https://www.businessinsider.com/james-cameron-bill-gates-ai-hard-write-science-fiction-2024-9?utm_source=chatgpt.com.

Barnes, B. and Koblin, J. (2023). Lengthy Strike Looks Likely for Hollywood Writers. *The New York Times*, B1-L.

Beer, J. (2023). Ryan Reynolds used ChatGPT to make a Mint Mobile ad, and the results were 'mildly terrifying.' *Fast Company.* January 10, 2023: https://www.fastcompany.com/90833253/ryan-reynolds-used-chatgpt-to-make-a-mint-mobile-ad-and-the-results-were-mildly-terrifying.

Beltran, M. A., Ruiz Mondragon, M. I. and Han, S. H. (2024, June). Comparative analysis of generative AI risks in the public sector. pp. 610–617. *In*: *Proceedings of the 25th Annual International Conference on Digital Government Research.*

Bengesi, S., El-Sayed, H., Sarker, M. K., Houkpati, Y., Irungu, J. and Oladunni, T. (2024). Advancements in generative AI: A comprehensive review of GANs, GPT, autoencoders, diffusion model, and transformers. *IEEE Access.*

Bratter, M. (2022). Matt Reeves' The Batman (2022). *UWIRE Text,* 1–1.

Bregler, C. (2007). Motion Capture Technology for Entertainment [In the Spotlight]. *IEEE Signal Processing Magazine*, 24: 160–158. https://doi.org/10.1109/MSP.2007.906023.

Campbell, V. and Campbell, V. (2016). Analytical frameworks: science, documentary and factual entertainment. *Science, Entertainment and Television Documentary*, 27–62.

Campbell, C. (2023). Ready or not, generative AI is here to stay: advertisers need more research to harness the benefits of AI technologies. *Journal of Advertising Research,* 63(3): 202–204.

Cassemere-Stanfield, A. (2023). *Touching Menace: Artificial Intelligence, Aesthetics, and The Intimacies of Synthetic Sense* (Doctoral dissertation, The University of Chicago).

Çelik, K. (2024). AI vs. Human in Screenwriting: Is AI the Future Screenwriter?. *Sakarya İletişim,* 4(1): 1–22.

Clark, S. (2024). Hitting the bricks with the creative class. *Labor,* 21(3): 77–83.

Cresswell, T. (2012). Capturing Mobility: Mobility and meaning in the photography of Eadweard Muybridge and Etienne-Jules Marey. pp. 57–83. *In*: *On the Move*. Routledge.

Crivelli, S. F. (2023). The spectacle of the moon conquest: how visual culture shaped Méliès' Le voyage dans la Lune and its anti-imperialist satire. *Early Popular Visual Culture,* 21(4): 407–433.

Davis, B. (2018). *Comic Book Movies*. Rutgers University Press.

Davies, G. (2019). David Beckham 'speaks' 9 languages for new campaign to end malaria. *ABC News.* April 9, 2019: https://abcnews.go.com/International/david-beckham-speaks-languages-campaign-end-malaria/story?id=62270227.

Dickson, W. K. L. and Dickson, A. (2000). *History of the Kinetograph, Kinetoscope, and Kinetophonograph.* The Museum of Modern Art.

Dwivedi, Y. K., Hughes, L., Ismagilova, E., Aarts, G., Coombs, C., Crick, T. et al. (2021). Artificial Intelligence (AI): Multidisciplinary perspectives on emerging challenges, opportunities, and agenda for research, practice and policy. *International Journal of Information Management,* 57: 101994.

Edwards, B. (2024). Toys "R" Us riles critics with "first-ever" AI-generated commercial using Sora. Ars Technica. June 26, 2024: https://arstechnica.com/information-technology/2024/06/toys-r-us-riles-critics-with-first-ever-ai-generated-commercial-using-sora/#gsc.tab=0.

Essien, E., Kalkat, A. and Lamichanne, S. (2024). Paradigm Shift of Visual Effects (Vfx) in Hollywood Apocalypses: Analyses of Advanced Cgi Artistry in Selected Films. *Available at SSRN 4925458.*

Etienne, V. (2021). Val Kilmer gets his voice back after throat cancer battle using AI technology: hear the results. *People.* April 19, 2021: https://people.com/movies/val-kilmer-gets-his-voice-back-after-throat-cancer-battle-using-ai-technology-hear-the-results/.

Farinacci, E. (2024). Towards a renewed understanding of screen and audiovisual education: a mapping of the relationship between AI and the film industry. *Scholé: rivista di educazione e studi culturali: LXII,* 1: 183–201.

Flavia, M. (2020). 25 Quentin Tarantino quotes to be ambitious with your creativity. *Goalcast.* September 9: 2020: https://www.goalcast.com/quentin-tarantino-quotes/.

Fleming, D. H. and Knee, A. (2020). The analogue strikes back: Star Wars, star authenticity, and cinematic anachronism. *Celebrity Studies,* 11(2): 205–220.

Foster, D. (2022). *Generative Deep Learning.* O'Reilly Media, Inc.

Frelik, P. (2022). Syd Mead (1933–2019). pp. 108–112. *In*: *Fifty Key Figures in Cyberpunk Culture.* Routledge.

Fteiha, B., Altai, R., Yaghi, M. and Zia, H. (2024). Revolutionizing video production: an AI-powered cameraman robot for quality content. *Engineering Proceedings,* 60(1): 19.

Gao, Y. and Liu, H. (2023). Artificial intelligence-enabled personalization in interactive marketing: a customer journey perspective. *Journal of Research in Interactive Marketing,* 17(5): 663–680.

Gardies, A. (2000). Cinema on show in the work of the Lumière brothers. *Echoes of Narcissus,* 2: 111.

Gerner, A. M. (2024). Playing with Arte (f) actors. pp. 9–43. *In*: *Diachronic Perspectives on Embodiment and Technology: Gestures and Artefacts.* Cham: Springer International Publishing.

Giacomin Da Silva, C. (2023). The intersection of AI and Art-Mapping the controversy around Generative AI tools as producers of artworks in the 2020s.

Grba, D. (2022). Deep else: A critical framework for AI art. *Digital,* 2(1): 1–32.

Green, B. R. (2023). *The Artist's Code: Technology and the Optimization of Creativity in Hollywood.* University of California, Los Angeles.

Guo, B., Shan, X. and Chung, J. (2024). A comparative study on the features and applications of AI tools-focus on PIKA labs and RUNWAY. *International Journal of Internet, Broadcasting and Communication,* 16(1): 86–91.

Hasan, M., Athrey, K. S., Khalid, A., Xie, D., Younessian, E. and Braskich, T. (2024). Applications of computer vision in entertainment and media industry. *Computer Vision: Challenges, Trends, and Opportunities,* 205.

Hasselgren, C. and Oprea, T. I. (2024). Artificial intelligence for drug discovery: Are we there yet?. *Annual Review of Pharmacology and Toxicology,* 64(1): 527–550.

Heaven, W. D. (2023). Welcome to the new surreal: How AI-generated video is changing film. *MIT Technology Review.* June 1, 2023: https://www.technologyreview.com/2023/06/01/1073858/surreal-ai-generative-video-changing-film/.

Huang, Y., Lv, S., Tseng, K. K., Tseng, P. J., Xie, X. and Lin, R. F. Y. (2023). Recent advances in artificial intelligence for video production system. *Enterprise Information Systems,* 17(11): 2246188.

Hutson, J. (2024, August 27). *Interview with Yonatan Dor of The Dor Brothers* [Interview].

Jin, Y., Sun, Z., Li, N., Xu, K., Jiang, H., Zhuang, N. et al. (2024). Pyramidal Flow Matching for Efficient Video Generative Modeling. *arXiv preprint arXiv:2410.05954.*

Karras, T., Aittala, M., Laine, S., Härkönen, E., Hellsten, J., Lehtinen, J. et al. (2021). Alias-free generative adversarial networks. *Advances in Neural Information Processing Systems,* 34: 852–863.

Katz, J. H. and Ellis, L. M. (2024). Dances with Avatar: How creators can reduce the novelty of their work to achieve more creative success. *Academy of Management Review,* (ja), amr-2022.

Kurop, N. (2003). State of the Art: Matte Painters—a Secret World of Illusion. *Metro Magazine: Media & Education Magazine,* (136): 154–157.

Kyriakoudes, K. J. (2024). Securing worker's futures: why replacing union workers with artificial intelligence in creative professions is the new subcontracting. *American University Law Review,* 73(5).

Kulesz, O. (2024). *Artificial Intelligence and International Cultural Relations: Challenges and Opportunities for Cross-Sector Collaboration.* Institut für Auslandsbeziehungen.

Labrecque, L. I., Peña, P. Y., Leonard, H. and Leger, R. (2024). Not all sunshine and rainbows: exploring the dark side of AI in interactive marketing. *Journal of Research in Interactive Marketing.*

Lanza, D. (2020). Brushstroke Cinema: The concept of matte painting in the Work of Peter Ellenshaw and Albert Whitlock. *Film History: An International Journal,* 32(2): 80–99.

Lee, W. and Masunaga, S. (2024). AI is supposed to be Hollywood's next big thing. What's taking so long? *Los Angeles Times.* October 17, 2024: https://www.latimes.com/entertainment-arts/business/story/2024-10-17/why-some-creatives-are-concerned-about-runways-deal-with-lionsgate.

Lee, Y. K., Park, Y. H. and Hahn, S. (2023). A portrait of emotion: Empowering self-expression through AI-generated art. *arXiv preprint arXiv:2304.13324.*

Li, K. (2024). Application of communication technology and neural network technology in film and television creativity and post-production. *International Journal of Communication Networks and Information Security,* 16(1): 228–240.

Lind, S. J. (2024). Can AI-Powered Avatars Replace Human Trainers? An Empirical Test of Synthetic Humanlike Spokesperson Applications. *Authorea Preprints.*

Liu, Y., Zhang, K., Li, Y., Yan, Z., Gao, C., Chen, R. et al. (2024). Sora: A review on background, technology, limitations, and opportunities of large vision models. *arXiv preprint arXiv:2402.17177.*

Loock, K. (2021). On the realist aesthetics of digital de-aging in contemporary Hollywood cinema. *Orbis Litterarum,* 76(4): 214–225.

Maselli, V. (2018). The evolution of stop-motion animation technique through 120 years of technological innovations. *International Journal of Literature and Arts,* 6: 54. https://doi.org/10.11648/J.IJLA.20180603.12.

Masood, M., Nawaz, M., Malik, K. M., Javed, A., Irtaza, A. and Malik, H. (2023). Deepfakes generation and detection: State-of-the-art, open challenges, countermeasures, and way forward. *Applied Intelligence,* 53(4): 3974–4026.

Matthews, J., Fastnedge, D. and Nairn, A. (2023). The future of advertising campaigns: The role of AI-generated images in advertising creative. *Journal of Pervasive Media,* 8(1): 29–49.

McAlister, A. R., Alhabash, S. and Yang, J. (2024). Artificial intelligence and ChatGPT: Exploring current and potential future roles in marketing education. *Journal of Marketing Communications,* 30(2): 166–187.

McNeese, T. (1989). Raiders of the lost art: Using the "painted word" in writing. (3): 34–36.

Meng, H., Xu, Q. and Sun, J. (2023). A study on the coupling of cinematic emotional narratives and social psychology: Makoto Shinkai's Animated Film Suzume. p. 01025. *In: SHS Web of Conferences* (Vol. 174). EDP Sciences.

Metz, R. (2024). AI Video Advances Bring It Closer to ChatGPT for Cinema. *Bloomberg.* March 20, 2024: https://www.bloomberg.com/news/newsletters/2024-03-20/chatgpt-for-movies-ai-video-tools-improve-technology-fast.

Michelle, C., Davis, C. H., Hardy, A. L. and Hight, C. (2017). *Fans, Blockbusterisation, and the Transformation of Cinematic Desire: Global Receptions of the Hobbit Film Trilogy.* Springer.

Mitchell, M. (2013). *Visual Effects for Film and Television.* Routledge.

Mugubi, J. G. and Manono, F. M. (2024). Optimizing creative processes with artificial intelligence (AI): Strategic imperatives for African filmmakers and graphic designers in the evolving digital landscape. *International Journal of Scholarly Practice,* 4(2): 28–40.

Murodillayev, B. (2024). The impact of visual effects on the cinema experience: a comprehensive analysis. *Art and Design Review,* 12(4): 238–249.

Nixon, L., Apostolidis, K., Apostolidis, E., Galanopoulos, D., Mezaris, V., Philipp, B. et al. (2024). AI and data-driven media analysis of TV content for optimised digital content marketing. *Multimedia Systems,* 30(1): 25.

Po, R., Yifan, W., Golyanik, V., Aberman, K., Barron, J. T., Bermano, A. et al. (2024, May). State of the art on diffusion models for visual computing. *In: Computer Graphics Forum* (Vol. 43, No. 2, p. e15063).

Pramanik, S. and Rai, S. K. (2023). AI take-over in literature and culture: truth, post-truth, and simulation. *Rupkatha Journal on Interdisciplinary Studies in Humanities,* 15(4).

Prince, S. (2011). *Digital Visual Effects in Cinema: The Seduction of Reality.* Rutgers University Press.

Purwanto, A., Utami, E. and Agustriawan, D. (2024, February). A comprehensive literature review on generative adversarial networks (GANs) for AI anime image generation. pp. 1–6. *In*: *2024 IEEE International Conference on Artificial Intelligence and Mechatronics Systems (AIMS)*. IEEE.

Rahman, M. D. and Ali, M. A. (2024). AI in video production: from script to screen. *Media and Al: Navigating,* 49.

Ramaul, L., Ritala, P. and Ruokonen, M. (2024). Creational and conversational AI affordances: How the new breed of chatbots are revolutionizing the knowledge industries. *Business Horizons.*

Reddy, V. S., Kathiravan, M. and Reddy, V. L. (2024). Revolutionizing animation: unleashing the power of artificial intelligence for cutting-edge visual effects in films. *Soft Computing,* 28(1): 749–763.

Rehak, B. (2007). The migration of forms: bullet time as microgenre. *Film Criticism,* 32(1): 26–48.

Rehak, B. (2018). 2. Used universes and immaculate realities: appropriation and authorship in the age of previzualization. pp. 71–106. *In*: *More Than Meets the Eye*. New York University Press.

Rojas, R. V. B. and Martínez-Cano, F. J. (eds.). (2024). *Revolutionizing Communication: The Role of Artificial Intelligence.* CRC Press.

Rosenstone, R. A. (2017). *History on Film/Film on History*. Routledge.

Sager, M. (2024). Watch the new Toys 'R' Us ad generated by AI. *Newsweek*. June 26, 2024: https://www.newsweek.com/toysrus-brand-film-ad-ai-generated-sona-1917645.

Sahin, F. and Soylemez, C. (eds.). (2024). *Globalized Consumer Insights in the Digital Era*. IGI Global.

Salam, M. A., Rayun, S. N., Islam, W., Hasan, R., Firmansyah, E. A. and Kalinaki, K. (2024). Consumer engagement: exploring deepfake applications in consumer marketing communication. pp. 397–421. *In*: *Navigating the World of Deepfake Technology*. IGI Global.

Sawant, O. (2024). *Visual Storytelling with Generative AI: A Practical Handbook for Modern Filmmakers and Content Creators*.

Shafer, E. (2024). Martin Scorsese doesn't think cinema is 'dying,' but 'transforming': 'it was never meant to be one thing.' *Variety.* February 20, 2024: https://variety.com/2024/film/festivals/martin-scorsese-cinema-dying-transforming-berlin-film-festival-1235916743/.

Sharakhina, L., Ilyina, I., Kaplun, D., Teor, T. and Kulibanova, V. (2023). AI technologies in the analysis of visual advertising messages: survey and application. *Journal of Marketing Analytics,* 1–24.

Sharf, Z. (2023). Christopher Nolen says AI dangers have been 'apparent for years'; press covered it more once chatbots threatened their jobs: 'suddenly it's a crisis.' Variety. June 20, 2023: https://variety.com/2023/film/news/christopher-nolan-artificial-intelligence-dangerous-filmmaking-1235649123/.

Sharma, V., Gupta, M., Kumar, A. and Mishra, D. (2021). Video processing using deep learning techniques: A systematic literature review. *IEEE Access,* 9: 139489–139507.

Shay, D. and Duncan, J. (1993). *The Making of Jurassic Park.* (No Title).

Shoaib, M. R., Wang, Z., Ahvanooey, M. T. and Zhao, J. (2023, November). Deepfakes, misinformation, and disinformation in the era of frontier AI, generative AI, and large AI models. pp. 1–7. *In*: *2023 International Conference on Computer and Applications (ICCA)*. IEEE.

Tao, Z., Lin, T. E., Chen, X., Li, H., Wu, Y., Li, Y. et al. (2024). A survey on self-evolution of large language models. *arXiv preprint arXiv:2404.14387.*

Thomas, S. (2024). AI and Actors: Ethical challenges, cultural narratives and industry pathways in synthetic media performance. *Emerging Media*, 27523543241289108.

Topçu, E. (2022). CGI (Computer Generated Imagery) Animasyon Tekniği ile Oluşturulan Olası Dünyaların Gerçekçiliğinin İncelenmesi. *İstanbul Aydın Üniversitesi Güzel Sanatlar Fakültesi Dergisi,* 8(15): 107–121.

Torun, A. (2025). Filmmaking and video art in the digital era. pp. 237–260. *In*: *Impact of Contemporary Technology on Art and Design*. IGI Global.

Venkatasawmy, R. (2012). *The Digitization of Cinematic Visual Effects: Hollywood's Coming of Age.* Lexington Books.

Wach, K., Duong, C. D., Ejdys, J., Kazlauskaitė, R., Korzynski, P., Mazurek, G. et al. (2023). The dark side of generative artificial intelligence: A critical analysis of controversies and risks of ChatGPT. *Entrepreneurial Business and Economics Review,* 11(2): 7–30.

Walley, J. (2022). *Expanded Cinema: Then and Now*. A Companion to Experimental Cinema, 59–83.

Weprin, A. (2024). James Cameron joins board of Stability AI in coup for tech firm. *The Hollywood Reporter.* September 24, 2024: https://www.hollywoodreporter.com/business/business-news/james-cameron-joins-board-ai-firm-stability-stable-diffusion-1236010034/.

Wong, D. C. and Williams, S. (2024). Artificial intelligence analysis of videos to augment clinical assessment: an overview. *Neural Regeneration Research,* 19(4): 717–718.

Wright, W. (2023). Coke launches 'Create Real Magic' AI art contest using GPT-4 and Dall-E 2. The Drum. March 20, 2023: https://www.thedrum.com/news/2023/03/20/coke-launches-create-real-magic-ai-art-contest-using-gpt-4-and-dall-e.

Yang, S., Yang, P., Chen, J., Ye, Q., Zhang, N. and Shen, X. (2023). Delay-optimized multi-user VR streaming via end-edge collaborative neural frame interpolation. *IEEE Transactions on Network Science and Engineering.*

Zhuang, Y. and Tang, S. (2021). Visual knowledge: an attempt to explore machine creativity. *Frontiers of Information Technology & Electronic Engineering,* 22(5): 619–624.

Chapter 2

The Art and Ethics of AI-Generated Filmmaking

This chapter investigates the evolving role of generative AI tools across the filmmaking process, from visual creation to editing and post-production. Beginning with key platforms like Stable Diffusion, MidJourney, and Runway, the chapter explores how these tools empower filmmakers to generate complex visuals, animate characters, and design settings with unprecedented ease. It addresses the evolving relationship between artists and AI, introducing the notion of computational creativity and questioning whether the tools can extend or merely simulate human creative expression. Additionally, the chapter examines aesthetics in AI-driven video art, prompting filmmakers to consider how these platforms reshapes traditional visual and narrative structures. Ethical concerns are also explored, focusing on authorship, copyright, and the social implications of using generative tools in film, especially regarding originality and ownership in AI-enhanced narratives. Finally, practical guidance is provided on incorporating such imagery, video sequences, music, and sound into a cohesive production, highlighting the balance between human agency and AI-generated elements. This chapter positions AI not as a replacement but as a collaborator, offering new dimensions to creative expression in the evolving world of filmmaking.

2.1 The Evolving Relationship Between Artists and AI

The interaction between creatives and generative AI (GAI) is redefining creativity and pushing the boundaries of artistic and filmmaking practice. As these generative platforms like Stable Diffusion, MidJourney, and Runway become increasingly integrated into the creative process, artists and filmmakers are challenged to rethink their roles, potentially embracing AI not merely as a tool but as a collaborative medium. The classical concept of *techne*—which encapsulates skill, craft, and knowledge—is being reshaped by the integration of GAI, offering new avenues for artistic experimentation, innovation, and expression. This evolution prompts creatives to explore how such tools influence their craft and workflow, leading to redefinitions of artistic *praxis*. Engaging with AI on this level requires an exploration of the ethical, social, and cultural implications it introduces to the realm of creative production (Hutson et al., 2023).

Despite traditional views that machines lack the capacity for true creativity, advances in computational art challenge this notion, presenting generative systems as both tools and active participants in the creative process. These platforms are now capable of generating poetry, music, visual art, and architectural designs, spanning a diverse array of artistic disciplines. This evolution positions AI as a dynamic force within the arts, one that goes beyond simple task automation to participate in creative decisions. Such capabilities prompt philosophical questions about the nature of creativity, as AI-generated works blur the line between human input and machine-driven processes. Consequently, this blurring redefines conventional ideas of authorship, as it becomes increasingly challenging to distinguish the contributions of the human creator from those of the algorithm. These dynamics also complicate audience engagement, as viewers must navigate interpretations of artworks influenced by both human and machine intent (Kharchenko et al., 2023).

The disruptive impact of these generative technologies lies not only in their multimodal capabilities across various media but also in the unprecedented speed of their adoption across creative industries. No previous technology has integrated so rapidly or provided such expansive tools for content generation and artistic experimentation. For example, OpenAI's ChatGPT set a historic record in 2022 by becoming the fastest-adopted technology, reaching one million users within five days and surpassing 100 million monthly active users by January 2023 (Gordon, 2023). GAI platforms like Claude, Stable Diffusion, and Runway have broadened the scope of creative possibilities, from generating music and visual art to drafting textual content, while also introducing new legal and ethical challenges. These technologies leverage machine learning models that generate content by identifying patterns within massive datasets (Feuerriegel et al., 2024). Unlike traditional tools like cameras or Photoshop, which require human input for capturing or editing existing content, GAI models are trained on extensive datasets of text, images, and other media, enabling them to learn structures, styles, and thematic elements. Consequently, the "generative" aspect of AI refers to the model's capability to create original content by merging learned elements in ways that emulate human creativity rather than merely modifying pre-existing material (Epstein et al., 2023).

For instance, platforms like Stable Diffusion (Fig. 2.1) and Runway generate visuals by using neural networks trained on vast image datasets to recognize and replicate visual patterns. When provided with a textual prompt, these models can synthesize images that align with the prompt's details. This process, powered by diffusion models or transformer architectures, produces visual or textual content that mirrors human-made pieces but originates from algorithmic synthesis rather than human-led design or manipulation (Moreno et al., 2023). Creators, through prompt engineering, can exercise control over the creative output, aligning the results more closely with their artistic intentions. Thus, the level of control over these outputs varies significantly, depending on the specificity and structure of the provided prompt—whether it is text, an image, a video, or an audio sample.

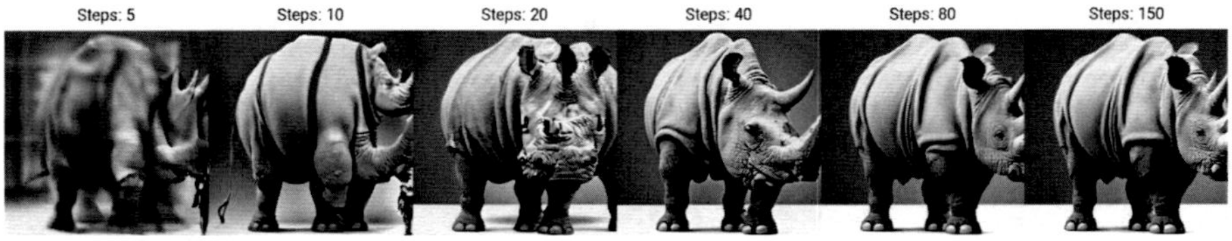

Figure 2.1 Rhino X Y Plot to Show Different Steps, Stable Diffusion, 2022 (Creative Commons License Zero).

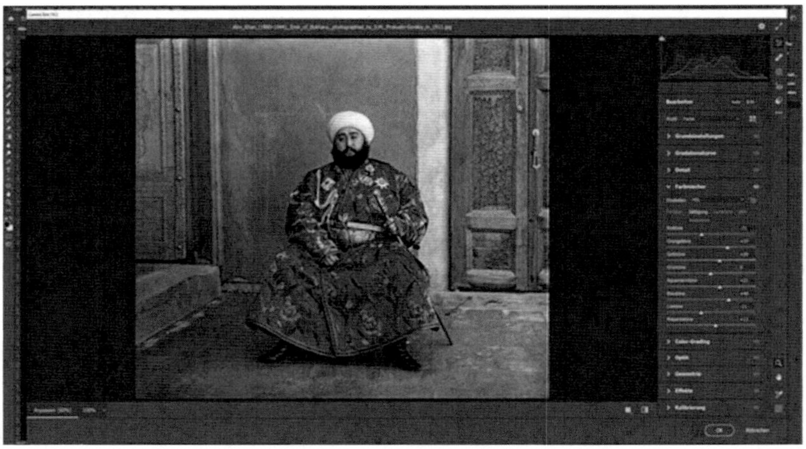

Figure 2.2 Screenshot Adobe Photoshop Camera RAW, 2022 (Creative Commons License Zero).

This range of human involvement distinguishes GAI from other creative tools, as it can function independently, generating outputs without necessitating a direct creative action from the user, such as taking a photograph or designing an image. Traditional tools like cameras and Photoshop are generally viewed as extensions of human creativity, relying on deliberate user actions and decisions. A camera, for instance, captures reality-based images, while Photoshop (Fig. 2.2) allows for enhancement or alteration. Even within these processes, human involvement remains essential; photography involves selecting a subject, framing, lighting, and adjusting lenses before capturing the image. Digital tools like Photoshop offer further control in post-production, allowing for a range of modifications, from minor adjustments like lighting changes and cropping to extensive alterations such as adding or removing subjects before finalizing the output for public display (Aaland, 2006; Kelby, 2020).

The degree of human involvement in the creative process is a highly debated aspect of creating with these platforms, particularly due to their capacity to interpret prompts and seemingly autonomously produce novel, algorithmically generated content. GAI utilizes its learned representations of various media elements without directly replicating specific pieces from its training data, enabling it to create entirely new outputs (Risi and Togelius, 2020). This ability to generate original content that can closely resemble human-made work presents both significant opportunities and considerable challenges. On one hand, it facilitates rapid and efficient content creation across diverse fields; on the other, it raises complex legal questions around authorship, copyright, and originality, as traditional copyright protections typically require direct and personal authorship—something often absent in AI-generated content (Abdikhakimov, 2023). This evolving landscape prompts us to examine deeper questions around creativity, particularly concerning artist intent, the limits of machine creative potential, and the extent to which AI can participate in what has traditionally been considered uniquely human expression.

The concept of creativity in the arts, particularly within filmmaking, has traditionally centered around the figure of a solitary, visionary creator—often the director in cinema—whose singular vision drives the project. Despite the perception of a single creative lead, filmmaking has always been a collaborative endeavor requiring input from actors, designers, editors, and technicians, each contributing to the realization of a film. Today, with the advent of generative AI, the process has shifted. While the role of generative systems can condense various creative functions into one platform, they simultaneously raise questions about whether this centralization of creative control dilutes or expands artistic expression. In the art world, grounded largely in physical media, the adoption of generative tools has brought about a significant culture shock. Conversely, film has always engaged with technology as a vehicle for creative ideas, where elements like editing and post-production are seen as integral components of the storytelling process, authentically enhancing human vision. With AI, the shift is not merely technological but structural, affecting the role of the artist in new ways.

As AI becomes increasingly sophisticated, it challenges the traditional relationship between artist and tool. In *The Creativity Code: Art and Innovation in the Age of AI*, Marcus du Sautoy argues that, while generative models can enhance creativity by drawing on vast databases and complex patterns, there remains a need to balance learning from established models with developing original ideas (du Sautoy, 2019). This view highlights both the benefits and the limitations of AI in creative work: it offers a powerful resource for pattern recognition and idea generation but risks encouraging a reliance on templates that may stifle originality. Similarly, Kurt (2018) contends that the technology, rather than replacing the artist, can serve as a partner, augmenting creative capacities and enabling artists to explore uncharted territories in their work. This chapter examines these evolving dynamics, positioning AI as an extension of creative practice, one that reshapes rather than replaces human ingenuity.

The concept of computational creativity—where algorithms contribute to the generation of art, music, literature, and design—has introduced a new frontier in artistic expression. The capabilities of generative tools extend beyond simple replication; they introduce unique elements that challenge traditional notions of originality and authorship. For instance, GAI algorithms can produce innovative patterns and ideas, offering a form of exploratory and transformational creativity that aids artists in expanding their artistic vision (Mazzone and Elgammal, 2019). However, the inability to capture the deeply personal and nuanced aspects of human creativity—such as emotional depth and lived experience—limits its function to that of an assistant rather than an originator. AI-generated art and video raises essential questions about the very nature of creativity: while it can push boundaries, it lacks the intentionality and cultural context that human artists inherently bring to their work.

As algorithms become integral to artistic production, issues of authorship and originality surface, especially as AI-generated content can closely resemble human-made art. Boden's (2010) model of creativity (Table 2.1), which classifies human creativity into exploratory, transformational, and imaginative types, offers a framework to understand the relationship between human and machine-generated art. In exploratory creativity, these tools can generate a wide array of variations, supporting human artists in their initial brainstorming phases. Similarly, transformational creativity allows them to offer new perspectives by generating outputs that challenge traditional norms. However, when it comes to imaginative creativity—rooted in personal and cultural contexts—the role of AI remains limited, as this form of creativity is inextricably tied to the human experience.

These three categories of creativity illustrate various approaches to generating new ideas and solutions, underscoring the complex nature of creativity across diverse disciplines. Rule-based creativity operates within defined rules and constraints, exploratory creativity is fueled by chance and experimentation, and intuitive creativity relies on intuition and insight. Recognizing these types is essential when examining the interaction between artists and generative AI. For example, AI excels primarily in the first type of creativity, utilizing established rules and algorithms to produce outputs. It also plays a significant role in the second, where exploration and randomness drive the generation of innovative ideas (Boden, 2010). Through its computational power, AI can support artists in the early stages of the creative process by identifying patterns, generating ideas, and suggesting unique directions, helping them expand the boundaries of their work (Hong and Curran, 2019). For instance, algorithms can analyze large datasets, detect correlations, and generate unexpected combinations that may not readily occur to human artists. This exploratory potential allows artists to investigate unconventional aesthetics and approaches, enhancing the exploratory dimension of creativity through data processing abilities (Mazzone and Elgammal, 2019).

However, AI struggles with the third type of creativity, which involves intuition, insight, and the deeply personal elements of human expression. The imaginative side of creativity, rooted in personal meaning, emotion, and intention, remains uniquely human. Artists and auteurs draw from personal and cultural experiences

Table 2.1 Boden's three categories of human creativity.

Type of Creativity	Description	Examples
Rule-Based Creativity	Relies on predefined methods and formulas to produce outcomes	Mathematics, engineering
Exploratory Creativity	Emphasizes experimentation and chance to discover new concepts	Visual arts, music, literature
Intuitive Creativity	Involves using instinct and deep understanding to develop ideas	Psychology, philosophy, social sciences

to create deeply resonant works, something beyond the current capacity of AI-assisted tools (Boden, 2010). Thus, while AI can support artists during exploratory and transformative stages, it is essential to acknowledge its limitations in capturing the human depth and emotion integral to intuitive creativity. Creatives can thus integrate the technology as a creative tool and partner, yet maintain their unique vision and the emotional nuances that only human creativity brings. Importantly, these creativity types are not mutually exclusive, as many artists embody a blend of them in practice. Different fields may emphasize certain types more than others, but recognizing creativity's diverse nature enables individuals and organizations to foster it holistically, using AI as an augmentation tool to enrich the creative process (Hong and Curran, 2019).

Kurt (2018) likewise suggests that generative capabilities can inspire artists with outcomes they may not have conceived independently, resulting in novel expressions that blend human and machine creativity. This fusion introduces questions about originality and ownership. For example, if an algorithm generates a new form based on training data, does the resulting work belong to the artist, the AI, or even the creators of the training data? This tension between human contribution and machine autonomy demands a reassessment of intellectual property laws and ethical standards within the art world (Sturm et al., 2019). As artists increasingly adopt AI, the ethical, social, and cultural ramifications of this technology come to the forefront. The blurred line between human and machine agency raises questions regarding the integrity of creative works. AI tools often draw from vast, uncredited datasets, sparking debate over the originality of AI-generated content and the rights of data contributors (Abdikhakimov, 2023). Additionally, the aesthetic qualities of AI-generated art—such as its precision, complexity, and adaptability—prompt artists to reconsider the value of traditional techniques. In a digital age where originality is increasingly challenged, the role of AI brings new perspectives to discussions on authorship, with some arguing that AI-driven art and video could disrupt traditional market dynamics by saturating the field with easily replicable digital works (Amankwah-Amoah et al., 2024).

Such concerns seem increasingly valid as the interplay between words and images within multimodal generative platforms exemplifies the field's rapid integration into artistic and commercial production, where text-to-text, text-to-image, and now text-to-video capabilities seamlessly blend across inputs and outputs (Akkus et al., 2023; Ananthanagu and Agarwal, 2023; Di Mitri et al., 2023). This shift marks a new phase in our historical understanding of knowledge, where AI-driven applications like Stable Diffusion, MidJourney, and ChatGPT not only produce novel content but challenge previous frameworks. Traditionally, knowledge in pre-modern societies was structured hierarchically—abstract principles held primacy, while specific facts were subordinate (Collet-Sabé, 2023). This paradigm shifted with the rise of nominalist philosophy, particularly William of Ockham's 14th-century rejection of universal concepts, which emphasized discrete entities over overarching truths, paving the way for empirical science that valued accumulating specific, observable facts (Orr, 2020).

As the scope of human knowledge grew, the demand for tools to organize and access this expanding data became crucial. Vannevar Bush's 1945 concept of the "memex" foreshadowed today's search engines, facilitating information retrieval and shifting reliance from personal memory to technological archives (Van Dijck, 2005). With tools like ChatGPT, the ability to summon vast knowledge instantly through simple prompts represents a culmination of this trend, echoing cultural shifts that began with Google's launch in 1998. This shift reshaped knowledge access, moving away from memorization toward an emphasis on rapid information retrieval (Säljö, 2010). However, despite their utility, generative AI systems lack the capacity to provide the abstract frameworks and ethical hierarchies humans use to interpret the world. While these models can access vast databases of information, they cannot independently construct insights related to morality, aesthetics, or deeper truths that require subjective human interpretation (Lazer et al., 2021). The responsibility for structuring this information meaningfully and discerning complex moral values remains uniquely human, underscoring the need for intentional guidance in an age of AI-assisted creativity.

Therefore, the rise of generative technologies have fundamentally altered how we create, allowing nearly instant access to visual and informational resources. Yet, while they serve as a powerful tool, the onus of interpreting and structuring content hierarchically still lies with human creators. This distinction is especially critical in generative art, where artists and designers play essential roles in shaping meaning and adding interpretative layers to AI outputs, maintaining a necessary balance between machine-generated content and human creativity.

A closer examination of computational creativity provides insights into how these generative systems intersect with traditional artistic practices, offering a framework to understand this evolving dynamic. Computational creativity, an interdisciplinary field, investigates the design and functioning of AI systems that display creativity or generate creative outputs. Through the integration of cognitive science, psychology, and computer science, computational creativity aims to model and simulate human creativity through algorithms and machine learning. This field examines creativity across idea generation, problem-solving, and artistic expression, not to replace human creativity but to extend it using computational means. The study of computational creativity includes developing algorithms that can autonomously produce innovative art, music, and literature, exploring the cognitive components of creativity like analogy and divergent thinking, which are foundational in human problem-solving and ideation.

Applications of computational creativity extend across art, music, literature, and beyond. Algorithms capable of generating ideas and recognizing patterns offer new possibilities for human-AI collaboration, pushing creative boundaries and reframing the role of AI as a partner in the artistic process. Events like the International Conference on Computational Creativity have been instrumental in exploring these ideas, providing a platform for researchers to present advances in AI-generated content across media, from visual art to literary and musical compositions. Several examples highlight the range of applications for computational creativity. Pinel and Varshney's (2014) work on AI-generated culinary recipes using big data showcases its reach into the culinary arts. Similarly, Duch's (2006) study of brain processes to develop naming algorithms demonstrates how computational systems can impact seemingly human-centric fields. Other researchers, like Jordanous (2016), argue for broader perspectives within computational creativity, emphasizing its multifaceted potential.

Specific studies demonstrate the practical implications of computational creativity. Znidarsic et al. (2016) explore conceptual blending in software composition, while Barreto et al. (2014) link procedural content generation to creative algorithms. Music, too, has seen advancements in AI-driven generation systems, with researchers like Cope (2015) and Carnovalini and Rodà (2020) examining how AI can innovate within compositional structures, enabling the creation of new music genres. Collectively, these studies illustrate computational creativity's potential across various domains. While computational algorithms can generate original works, they act as supplements to human artistry, augmenting rather than replacing it. Artists contribute a personal, emotional, and contextual richness that algorithms cannot replicate, ensuring that even in an era of sophisticated AI, the human element in creativity remains irreplaceable.

2.2 Exploring the New Aesthetics of AI

In the domain of AI-generated art, a key question emerges: can traditional aesthetic theories adequately interpret the distinctive qualities of AI-driven works, or does this new form demand fresh perspectives? The tension between existing theories and the need for new approaches defines much of the current discourse around AI aesthetics. As algorithms increasingly influence the creative process, it is crucial to reassess and expand aesthetic frameworks to fully appreciate the transformative potential AI brings to art. This exploration invites a rethinking of traditional aesthetic norms, aiming to capture the nuances that AI introduces into the artistic landscape and the novel experiences it enables (Akkus et al., 2023; Ananthanagu and Agarwal, 2023; Di Mitri et al., 2023).

To understand the influence of AI on art, one must first consider the aesthetics of the field, which has evolved over centuries. Historically, aesthetics has examined beauty, taste, and the role of art in human experience, beginning with philosophical underpinnings from the Enlightenment. During this era, thinkers like Immanuel Kant (1724–1804) shifted focus from universal principles to subjective experience, proposing that art could be appreciated for its own sake, independent of function or purpose (Dwivedi, 2021; Lehman, 2020). The Kantian concept of "disinterested pleasure" fundamentally reshaped aesthetic theory by proposing that true appreciation of beauty occurs when we contemplate an object without concern for its practical utility or personal benefit. This revolutionary framework suggests that authentic aesthetic experiences arise when we engage with art purely for its contemplative value, free from considerations of usefulness or personal gain. Kant argued that this type of aesthetic appreciation has a universal quality—when we judge something as beautiful in this disinterested way, we implicitly suggest that others should find it beautiful too, pointing to a shared human capacity for aesthetic experience that transcends individual preferences or practical concerns.

This Kantian framework provides a particularly valuable lens for understanding and evaluating AI-generated art, shifting our focus from questions of technical origin to the fundamental nature of the aesthetic experience itself. Rather than becoming entangled in debates about whether AI can be "truly creative" or whether machine-generated art is "authentic," Kant's perspective invites us to evaluate AI art based on its capacity to create genuine aesthetic experiences that transcend practical concerns. This approach suggests that the value of AI art lies not in its algorithmic sophistication or technical achievement, but in its ability to evoke universal aesthetic responses and create moments of pure contemplative delight. Just as we can appreciate the beauty of a natural landscape without understanding its geological formation, we can engage with AI art purely on the basis of its experiential impact, independent of its technological origins.

This Kantian framework provides a valuable lens for understanding AI-generated art, but becomes even more intriguing when considered alongside the critique of authorship by Michel Foucault (1926–1984). While the Kantian perspective invites us to evaluate art based on its capacity to create genuine aesthetic experiences that transcend practical concerns, Foucault's "What is an Author?" (1969) challenges us to move beyond traditional notions of singular creative agency. Foucault fundamentally challenges the romantic notion of the author as a singular genius who imbues works with definitive meaning, instead arguing that the "author function" serves as a cultural mechanism for constraining and controlling how texts circulate and are interpreted within society. He suggests that our modern conception of authorship, far from being a natural or eternal truth, emerged from specific historical and social conditions during the Enlightenment period, when various legal, economic, and cultural forces coalesced to create what we now think of as literary ownership and creative attribution.

By identifying authorship as a contingent social construct rather than an essential truth, Foucault opens up the possibility of more fluid and popular approaches to textual interpretation, where meaning can emerge from the interaction between text and reader without being constrained by appeals to authorial intent or biography—a perspective that becomes particularly relevant in the age of AI-generated content, where traditional notions of singular human authorship are increasingly challenged by collaborative human-machine creative processes. The emergence of AI art systems fundamentally disrupts the conventional author-work relationship, creating what might be called a "distributed creativity" where the final artistic output emerges from a complex interplay between human intention, algorithmic processes, and training data. This collaborative dynamic between human and machine challenges us to reconsider not just how we attribute creativity, but whether our traditional frameworks for understanding artistic creation remain relevant in an age of AI-assisted art.

The synthesis of Kantian aesthetics and Foucauldian critique offers a rich theoretical foundation for appreciating AI art on its own terms. Rather than becoming entangled in questions of authenticity or trying to parse exact proportions of human versus machine contribution, we can focus on the capacity of the artwork to generate meaningful aesthetic experiences while acknowledging the complex network of agencies involved in its creation. This approach suggests that the value of AI art lies not in traditional notions of originality or in technical sophistication, but in its ability to evoke universal aesthetic responses while simultaneously challenging our assumptions about creative agency. The result is a more nuanced understanding of artistic creation that recognizes both the importance of the aesthetic experience itself and the complex collaborative nature of contemporary artistic practice, where the boundaries between human and machine creativity become increasingly fluid and intertwined.

These redefinitions of creativity fundamentally reshape how we evaluate AI-generated art, moving beyond traditional metrics of originality and technical skill to consider more nuanced aspects of creative expression. One such example comes by way of a study by Ahmed (2022) that demonstrates that AI should be understood not merely as a tool but as a design element capable of embodying intangible human characteristics like emotions and memory. This perspective suggests that AI systems can function as creative collaborators that enhance and amplify human artistic vision, rather than simply executing predetermined commands. Furthermore, the technology's ability to engage with and respond to human emotions opens up new possibilities for interactive and responsive artworks that blur the boundaries between creator, medium, and audience.

The role of AI in shaping interactive experiences becomes particularly significant when considering immersive media installations, where the technology can create dynamic, responsive environments that evolve based on audience engagement (Ahmed, 2022). This capacity for real-time adaptation and response fundamentally transforms the relationship between artwork and viewer, creating experiences that are both personally meaningful and collectively shared. Rather than replacing human creativity, AI augments it by enabling new forms of artistic expression that were previously impossible, while simultaneously challenging traditional notions of authorship and creative agency. This reorientation suggests that the true value of AI in art lies not in its ability to generate content autonomously, but in its capacity to facilitate deeper, more meaningful connections between artists, audiences, and the creative process itself, ultimately enriching the human experience of art rather than diminishing it.

The debate over creative autonomy in machines continues to evolve as artificial intelligence systems become increasingly sophisticated in their artistic capabilities. Csikszentmihályi's (1988) model of creativity provides a valuable framework for examining this issue, outlining three essential components: a knowledge domain, an agent introducing novelty, and a panel of experts for validation. When applied to AI systems, this model raises fundamental questions about whether machines can truly operate within established creative domains while introducing meaningful novelty. Jennings (2010) builds upon this foundation by proposing specific criteria for machine creative autonomy, emphasizing that AI systems must develop independent assessment capabilities, initiate changes without human prompting, and implement non-random evaluation processes.

The transition of AI systems from creative tools to potential creators represents a significant shift in how society conceptualizes artistic production. These systems must demonstrate a substantial degree of independence from human influence to achieve what Jennings (2010) defines as "creative autonomy." This independence extends beyond mere technical capability to encompass sophisticated decision-making processes about artistic choices and aesthetic outcomes. The ability of AI to evaluate the quality and impact of its creations independently marks a crucial step toward establishing genuine creative agency, though the parameters of such agency remain subject to ongoing debate within the artistic and technological communities.

The relationship between creativity and external validation emerges as a critical consideration in evaluating AI-generated art. Ajani (2022) emphasizes that creative legitimacy inherently requires recognition and validation from domain experts, suggesting that AI creativity remains fundamentally connected to human assessment frameworks. This dependency on human judgment indicates that even as AI systems develop greater autonomous capabilities, the ultimate determination of artistic merit continues to reside within human experiential and evaluative domains. The interconnected nature of creative validation raises important questions about the potential limitations and opportunities for AI systems operating within established artistic contexts.

Cheng (2022) introduces an alternative perspective by proposing new frameworks for understanding AI-generated art that acknowledge the unique characteristics distinguishing it from traditional human-created works. By applying Schema Theory to the analysis of AI art, Cheng demonstrates how audiences engage with these works through the lens of existing artistic knowledge while simultaneously encountering novel aesthetic experiences that challenge conventional interpretative frameworks. This theoretical approach suggests that the appreciation of AI art requires a dynamic understanding that bridges traditional artistic schemas with emerging forms of creative expression, potentially leading to new paradigms for evaluating and experiencing art in the digital age. The intersection of human perception and machine-generated creativity creates opportunities for expanding the boundaries of artistic appreciation while maintaining connections to established aesthetic traditions.

Turning specifically to the nascent visual aesthetics of early tools, the emergence of AI art has introduced a distinctive visual vocabulary that sets it apart from traditional artistic mediums. Early AI-generated images demonstrated characteristic distortions and surreal elements, from anatomical irregularities to uncanny facial expressions, which became defining features of this nascent art form (Mori, 2012). In the realm of AI video generation, these peculiarities extended into the temporal dimension, creating sequences that often struggled with maintaining consistent motion and coherent narrative flow. The resulting aesthetic closely aligned with the concept of the "uncanny valley," producing visuals that simultaneously attracted and unsettled viewers through an almost-but-not-quite-human quality.

The technical limitations of early video systems manifested in specific visual artifacts that became hallmarks of the medium. Motion often appeared fluid yet subtly wrong, with human figures moving in ways that defied natural physics. Facial features might shift or morph between frames, creating an unsettling sense of instability (Figs. 1.5, 1.6, 1.7). These characteristics drew parallels to historical photographic experiments, particularly Victorian spirit photography, where technical limitations and artistic intention combined to create images that existed at the intersection of the real and the supernatural. These early AI videos frequently exhibited what critics described as a "fever-dream" quality, where familiar elements combined in impossible ways, creating a new form of digital surrealism.

The evolution of AI video technology has gradually refined these initial aesthetic qualities, moving toward more controlled and photorealistic output. However, the impact of early AI aesthetics continues to influence contemporary digital art practices. These pioneering works established a visual language that resonated deeply with digital culture, celebrating rather than concealing the artifacts of machine learning processes. The glitches, distortions, and uncanny elements that marked early AI videos have become recognized as legitimate artistic elements, similar to how the grain of film or the pixelation of early digital art became aesthetic choices rather than mere technical limitations.

This transition from technical limitation to artistic choice marks a significant moment in the development of AI as a creative medium. As the technology advances, artists and creators increasingly engage with these early aesthetic markers intentionally, incorporating them into works that explore the boundaries between human and machine creativity. These stylistic elements serve as reminders of AI's rapid evolution while simultaneously questioning traditional notions of artistic intent and aesthetic value. The preservation of these early aesthetic qualities, even as technology progresses beyond the initial limitations that created them, suggests a maturing artistic medium that acknowledges and builds upon its historical development while moving toward new forms of creative expression.

The ongoing refinement of these video tools continues to expand the possibilities for creative expression while maintaining connections to the medium's distinctive early aesthetic heritage. This evolution raises fundamental questions about the relationship between technical capability and artistic intent, suggesting that as AI technology becomes more sophisticated, the choice to embrace or reject these early aesthetic markers becomes an increasingly significant artistic decision. The deliberate incorporation or exclusion of these characteristics reflects broader discussions about authenticity, artistic intent, and the role of imperfection in digital art creation.

2.3 Ethics, Authorship, and Social Implications

The integration of generative systems into artistic creation presents a landscape of complex ethical considerations that extend far beyond traditional creative practices. The ability of generative systems to produce artwork by processing vast datasets of human-created content raises fundamental questions about originality, attribution, and fair compensation within creative industries. Recent developments in ML tools have sparked heated debates around copyright protection, with cases emerging where synthetic media outputs closely mirror existing artistic styles without explicit permission or compensation (Ren et al., 2024). The rapid advancement of generative technologies has outpaced current legal frameworks, creating uncertainty around intellectual property rights and proper attribution mechanisms (Lescrauwaet et al., 2022). These concerns become particularly acute when considering how automated systems parse and recombine elements from training data, potentially incorporating copyrighted material into new works without clear mechanisms for acknowledgment or compensation (Quang, 2021).

The transformation of artistic practice through computational creativity challenges established notions of authorship and creative agency. As ML systems become more sophisticated in generating visual content, music, and literature, the traditional concept of a sole human creator becomes increasingly complex to define and defend. Contemporary artists often function as curators and directors of artificially intelligent systems, guiding and refining machine-generated outputs rather than creating every element from scratch. This shift in creative methodology requires new frameworks for understanding collaborative creation between human artists and computational systems. The emergence of hybrid creative processes, where synthetic media tools augment and enhance human creativity, suggests a future where authorship becomes more distributed and multifaceted than traditional artistic paradigms have allowed.

The accessibility of creative tools through generative technologies introduces significant social implications that reshape the landscape of cultural production. ML platforms have made sophisticated creative capabilities accessible to individuals without formal artistic training, potentially disrupting established hierarchies within creative industries. However, this opening up of the industry also raises concerns about the displacement of human creatives and the potential homogenization of creative expression through widely available algorithmic tools. The environmental impact of computational creativity presents another crucial consideration, as the training and operation of large synthetic media models consume substantial computational resources and energy. These long-standing ecological concerns become particularly relevant when considering the proliferation of digital art markets and blockchain-based art platforms, which often require significant computational infrastructure (Rykiel, 1989).

Questions of bias and representation within generative systems emerge as critical ethical considerations that demand careful examination and ongoing dialogue. The training data used to develop synthetic media tools often reflects existing societal biases, potentially perpetuating or amplifying problematic representations in generated content. Cultural appropriation through machine learning systems raises complex questions about the responsible use of artistic traditions and cultural expressions in training data (Shahbazi et al., 2023). The development of ethical frameworks for computational creativity must address these concerns while fostering inclusive and equitable creative practices. As generative technologies continue to evolve, establishing guidelines for responsible development and deployment becomes essential for ensuring that synthetic media tools contribute positively to cultural expression while respecting diverse artistic traditions and communities.

The widespread accessibility and quick adoption rate of new AI-driven art tools owes much to their browser-based interfaces, which have significantly reduced dependency on proprietary software and made these platforms more approachable for users without specialized technical skills. This shift has effectively lowered barriers to entry, promoting broad adoption and facilitating creative experimentation. The ease of access has allowed a diverse range of users, from hobbyists to professionals, to engage with these tools, sparking a surge of interest in the artistic potential of AI technologies. The popularity of these platforms speaks to a desire among users to explore new creative horizons and push the boundaries of digital art. However, this rapid public embrace has also intensified debates within the art community (Ansari, 2022; Murphy, 2022; Hazucha, 2022).

Amid this enthusiasm, traditional artists and designers have voiced strong concerns regarding copyright, attribution, and the legitimacy of AI-generated art as a new genre. The rise of AI-generated artwork has led to questions about intellectual property rights and the status of AI as a co-creator. This tension underscores a growing cultural shift: while AI tools widen access to art-making by enabling non-artists to produce high-quality visuals, they also challenge the value and recognition of traditionally trained artists. These concerns highlight the divide between those who view AI as a tool for creative expansion and those who fear it undermines the authenticity and labor-intensive aspects of conventional art. The legal landscape remains uncertain, with institutions like the U.S. Copyright Office struggling to adapt existing frameworks to accommodate AI-assisted creations.

A recent lawsuit brought by Getty Images against Stability AI, creators of Stable Diffusion, has further highlighted these legal and ethical dilemmas. Filed on January 17, 2023, Getty Images' claim alleged that Stability AI had "unlawfully copied and processed millions of copyright-protected images" to train its software, igniting a debate over whether this practice constitutes fair use or infringement (Vincent, 2023). Stability AI defended its actions by arguing that scraping publicly available images for training data is permissible under U.S. fair use doctrine. In contrast, rights holders like Getty Images assert that this usage disregards the rights of artists and compromises the integrity of copyrighted material. Getty Images CEO Craig Peters likened the current legal ambiguities surrounding AI to the early controversies of digital music, expressing hope that this legal action would bring clarity to the issue of intellectual property in generative media. As the press release states,

> Getty Images believes artificial intelligence has the potential to stimulate creative endeavors. Accordingly, Getty Images provided licenses to leading technology innovators for purposes related to training artificial intelligence systems in a manner that respects personal and intellectual property rights. Stability AI did not seek any such license from Getty Images and instead, we believe, chose to ignore viable licensing options and long-standing legal protections in pursuit of their stand-alone commercial interests (Vincent, 2023).

The Getty lawsuit exemplifies the broader legal challenges posed by generative AI technologies, particularly as they intersect with questions of copyright. Unlike traditional tools that assist artists in directly manipulating media, generative AI platforms often rely on vast datasets of existing art, raising ethical concerns about originality and attribution. Many artists argue that AI's reliance on such datasets to generate new images undermines the concept of "original" art, as AI outputs may incorporate elements from countless copyrighted works without the creators' consent.

This reliance on prior artworks for algorithmic training thus raises fundamental questions about how society values and protects creative labor in the digital age.

The implications of this legal debate extend beyond individual cases and have broader ramifications for AI's role in creative industries. While companies like Stability AI argue for the fair use of data scraping, opponents highlight the need to protect artist intellectual property and compensation. With generative technologies continuing to evolve and permeate cultural production, the urgency to establish clear guidelines and frameworks for fair and ethical use intensifies. Ultimately, these legal precedents will play a crucial role in shaping the future landscape of AI-generated art, influencing both the rights of artists and the responsibilities of AI developers as they navigate the complex intersection of creativity, technology, and intellectual property.

The disruption of established copyright frameworks stems largely from their historical emphasis on human authorship as fundamental to intellectual property protections (Ploman and Hamilton, 2024). Copyright law, traditionally built upon the notions of originality and intentional human creativity, now struggles to adapt to the inclusion of AI-generated works—creations generated by data-driven algorithms rather than by conscious human authorship. A notable example illustrating these complexities is the ongoing case *Andersen et al. v. Stability AI Ltd.* (August 12, 2024), where a group of artists has alleged that Stability AI and other GAI companies infringed upon copyrights by training their models with billions of online images sourced without explicit permission. This case, currently under consideration in the California courts, raises essential questions about the legality of training AI models on datasets containing copyrighted images, and whether such datasets can themselves be classified as infringing works. Additionally, the case probes the issue of assigning culpability to companies that develop and deploy these generative AI tools (No. 23-cv-00201-WHO (N.D. Cal. Aug. 12, 2024)). This example underscores the limitations of current copyright law, which was not originally designed to address the intricacies of data-driven, AI-based content generation. As the courts weigh whether utilizing copyrighted images to "train" AI models constitutes infringement—or if, as argued by these companies, the process qualifies as "fair use" due to non-replicative output—the necessity for updated legal frameworks becomes increasingly evident (Spica, 2024).

The reliance on vast datasets of human-created content for AI model training raises important questions about the role of human creators in the resulting output, particularly concerning the definition of "authorship" across different media. A central issue lies in copyright's fundamental requirement for originality, implying an intentional, creative contribution by a human author. Historically, U.S. copyright law, as outlined in 17 U.S.C. § 102, mandates a "human author" for a work to qualify for copyright, a standard that AI-generated works challenge since these outputs often lack direct human authorship (Abbott and Rothman, 2023). Legal precedents such as *Allen v. U.S. Copyright Office* (2024) (1:24-cv-2665) illustrate the challenges in applying traditional standards to AI-assisted creations. In this case, artist Jason Allen sought copyright for his AI-assisted artwork *Théâtre D'opéra Spatial* (2022) (Fig. 1.4), created using the GAI tool Midjourney. The U.S. Copyright Office denied

Allen's application, citing insufficient human authorship—a foundational criterion for copyright protection. Allen contended that his careful prompting constituted a form of creative authorship, thus establishing him as the rightful creator. However, the Copyright Office maintained that copyright protections necessitate human creativity rather than merely directing an algorithm, thereby reinforcing the importance of direct human intervention in copyright eligibility (Brittain, 2024).

This decision aligns with other recent rulings, including *Thaler v. Perlmutter*, in which the court determined that "autonomous creations" by AI cannot qualify for copyright protection due to the absence of human authorship. The dispute began in August 2019 with *A Recent Entrance to Paradise* (Fig. 2.3), an image that computer scientist Stephen Thaler claimed was autonomously generated by his Creativity Machine. Thaler argued that, as the machine's owner, he should hold the copyright. However, the court denied this claim, emphasizing that copyright law necessitates a human author (Escalante-De Mattei, 2024). This case highlights the challenges in categorizing AI-generated works under conventional copyright frameworks and underscores the need for clearer guidelines on the level of human involvement required for copyright eligibility as AI becomes more integrated into creative processes. Additionally, lawmakers' deeper understanding of how creators use these tools could foster a more nuanced interpretation of "autonomous creations" (Mathur, 2023).

Another instructive example is the case of *Zarya of the Dawn* (2022) (Fig. 2.4), which illustrates the evolving stance of copyright law on AI-generated content and the critical role of human input in determining copyright eligibility. This AI-assisted graphic novel by Kris Kashtanova initially faced rejection by the U.S. Copyright Office, which argued that the images generated through MidJourney lacked the human creativity required for copyright protection. However, the Office granted copyright for the human-authored narrative and the comic's overall structure, marking a distinction between purely AI-generated components and human-crafted elements. The Copyright Office's decision reaffirms that copyright law under 17 U.S.C. § 102 centers on original human authorship, which AI cannot fulfill independently. This case reflects the Office's commitment to positioning human creativity at the heart of copyright protection, emphasizing the necessity of meaningful human input for legal recognition (Klukosky and Kohel, 2024).

The current state of copyright law, alongside the powerful capabilities of GAI tools like MidJourney and DALL-E, underscores the importance of human-guided prompts and creative direction in producing copyrightable works. These tools operate by synthesizing outputs from large datasets using algorithmic patterns, legally positioning them as extensions of human intention rather than as independent creators. This legal approach aligns with past copyright precedents, such as *Feist Publications* and *Meshwerks*, which established that originality arises from human creative effort rather than from mechanical reproduction. Consequently, the level of human input in guiding AI outputs is essential for determining copyright eligibility, reinforcing that human creativity remains central in AI-assisted works (Militsyna, 2023).

Establishing copyright for AI-generated works hinges on specific types of human input that demonstrate originality and creative decision-making, both

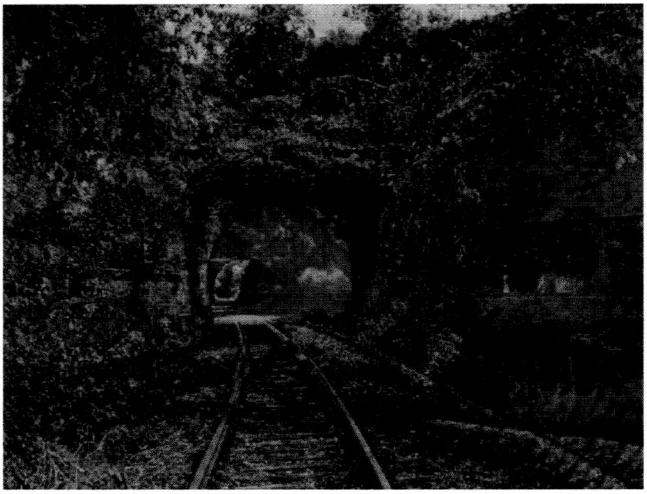

Figure 2.3 Stephen Thaler, *A Recent Entrance to Paradise*, 2023. Creativity Machine (Creative Commons License Zero).

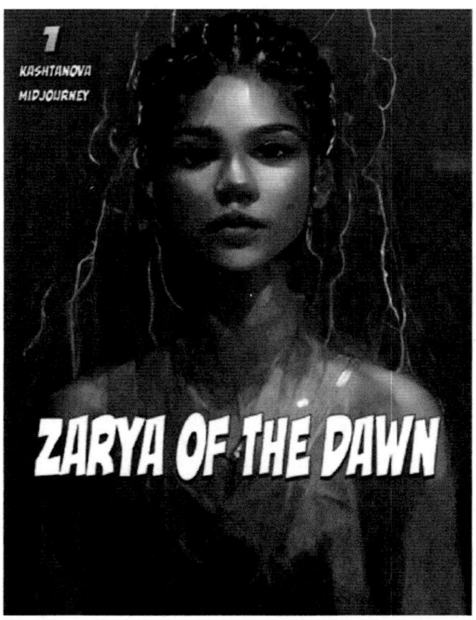

Figure 2.4 Kris Kashtanova, *Zarya of the Dawn*, Midjourney, 2022

essential components of copyright law (Table 2.2). One critical element of human involvement in creating AI-generated art is the design of prompts. Through detailed instructions that define style, composition, and thematic focus, the human creator provides a conceptual framework that the AI then brings to life. This process reflects the principles in *Mannion v. Coors Brewing Co.*, where the court recognized that creative choices—such as lighting, framing, and angle—imbue photographs with the originality required for copyright protection. Likewise, a carefully crafted prompt encapsulates the human creator's unique artistic vision, positioning the AI-generated

Table 2.2 Human contributions to AI-generated art and copyright implications.

Type of Human Input	Description	Legal Parallel
Creative Prompt Design	Human-designed prompts specifying style, composition, and theme establish a conceptual framework, reflecting human creativity and intention.	*Mannion v. Coors Brewing Co.*: Recognized that creative choices, like lighting and framing, add originality to a work.
Selection and Curation of Outputs	Choosing specific outputs from multiple AI-generated options demonstrates subjective choice, akin to selecting final shots in photography.	*Garcia v. Google, Inc.*: Emphasized the importance of control over final work for authorship.
Post-Processing and Refinement	Modifying AI-generated images by adjusting colors, altering compositions, or adding elements creates a unique creative layer, transforming the output into a derivative work.	*Meshwerks, Inc. v. Toyota Motor Sales, U.S.A., Inc.*: Recognized that significant alterations can impart originality to reproductions.
Conceptual Framework and Artistic Intent	Thematic and conceptual decisions by human creators shape the essence of the final work, signifying originality.	*Feist Publications, Inc. v. Rural Telephone Service Co.*: Reinforced that originality arises from creativity and intent.
Human Authorship Threshold	The degree of human involvement, including textual and narrative elements, impacts copyright eligibility, especially when AI-generated content alone is insufficient.	*Zarya of the Dawn* decision by the U.S. Copyright Office: Granted copyright for the human-authored narrative, not for AI-generated images lacking human input.
Derivative Work Creation	Transforming AI outputs through creative modifications aligns with principles of originality, allowing copyright for derivative works with significant human input.	17 U.S.C. § 101: Defines derivative works as those transforming or building upon preexisting materials through creative additions.

output for copyright eligibility by showcasing the user's contribution to the work's artistic direction (Burylo, 2022).

The act of selecting and curating outputs from multiple AI-generated options serves as a critical marker of human authorship. Just as photographers review numerous shots to identify those that best reflect their creative vision, choosing specific AI outputs from a range of generated possibilities involves a high degree of artistic discretion. This process of curatorial decision-making not only shapes the final work but also embeds the creator's unique artistic intent, reinforcing a foundation for copyright claims. Through the selection of one version over another, creators exercise subjective judgment, aligning the result with a particular vision and underscoring the vital role of human choice in the creative process (Wan and Lu, 2021).

In addition, post-processing and refinement play a significant role in establishing copyright eligibility for AI-assisted creations. When creators enhance AI-generated images by refining colors, adjusting compositions, or adding elements, they introduce new creative layers that distinguish the work from a straightforward automated output. This form of human input parallels the precedent in *Meshwerks*, where substantial modifications imbued digital reproductions with originality. Within the realm of AI, extensive post-processing converts generated content into a derivative work, reflecting human creativity and justifying copyright protection through the creator's transformative intervention (Geiger, 2023).

Furthermore, the overarching conceptual framework and artistic intent guiding AI tool use are essential in determining authorship. Whether employing AI to examine social issues, convey targeted messages, or reflect specific aesthetic philosophies, it is the human creator's vision that shapes the work's core essence. This aligns with insights from *Mannion*, where the court acknowledged the photographer's creative choices as foundational to copyrightability. In AI-generated art, the creator's intent and philosophical direction affirm the originality requirement in copyright law, emphasizing that AI functions as an instrument of the human artist's broader objectives (Kasap, 2021).

Together, these facets of human involvement reveal a spectrum of creativity that differentiates AI-assisted works from purely algorithmic outputs. From carefully crafted prompts and selective curation to significant post-processing and thematic intent, these elements collectively highlight that meaningful human creativity is essential for copyright claims in the age of AI. As AI technologies advance, the consistent need for substantial human contribution will likely remain central to copyright eligibility, preserving a balance between innovation in technology and the recognition of human authorship. These varied forms of human involvement indicate that AI should be seen as a tool that enhances, rather than replaces, human creativity. The *Compendium of U.S. Copyright Office Practices* underscores this notion, specifying that significant human intervention is necessary to satisfy the originality standard in copyright law. This guideline supports the Copyright Office's position in cases such as *Zarya of the Dawn*, aligning with 17 U.S.C. § 102, which emphasizes the need for human authorship. Moving forward, copyright considerations for AI-generated content are expected to focus on whether human input is sufficiently creative to fulfill the originality requirement (Geng, 2023).

For artists leveraging AI tools, adhering to specific practices can help establish copyright eligibility. First, crafting detailed prompts that demonstrate originality and creative intent can highlight human authorship. Second, selecting and curating specific AI-generated outputs reflects subjective artistic judgment, adding weight to copyright claims. Third, undertaking substantial post-processing work to transform the AI output into a derivative creation underscores the human contribution. Finally, maintaining a clear conceptual framework directed by the vision of the creative reinforces the notion of AI as a supportive tool rather than an independent creator, strengthening the copyright eligibility of the work. Considering these elements, potential updates to copyright law could introduce a category that formally acknowledges AI-assisted works as collaborative creations. The approach respects the core of human creativity while accommodating machine capacities, enabling copyright law to remain effective and inclusive in a creative environment increasingly shaped by AI. Such a framework would ensure that the creative contributions of human artists are protected, even as AI becomes more integral to artistic processes.

2.4 Technical Challenges and Opportunities

To develop copyright frameworks that effectively recognize human contributions in AI-assisted works, it is essential to incorporate practical recommendations grounded in both legal standards and artistic processes. Establishing a model that acknowledges human creative involvement while adapting to the distinct nature of generative AI can help ensure that AI-assisted works are legally protected and culturally valued (Makam and Dutta, 2023). In traditional cinematic production, creativity is a collaborative effort involving screenwriters, directors, actors, set designers, and editors, among others. Each role contributes uniquely to the final product, and copyright protections reflect this collective input. Introducing structured guidelines for AI-generated content in filmmaking can create a similar framework, ensuring that the artistic contributions of humans are both credited and protected. Before AI integration, traditional cinematic workflows followed specific steps that clarified creative input and authorship, laying a foundation for re-envisioning this process with generative tools (Harrison, 2003).

The filmmaking process has traditionally involved a series of meticulously planned stages, each requiring specialized roles and significant resources. From the initial scriptwriting and concept development to final post-production and distribution, cinematic creation has long been a labor-intensive endeavor that relies on the expertise of writers, directors, designers, and editors. Each phase contributes to the creative and technical foundation of the final film, with tasks like scriptwriting, storyboarding, set design, and post-production often taking months, if not years, to complete. Traditional production workflows are centered around human creativity and hands-on technical work, with each stage building upon the previous to ensure that the director's vision is meticulously captured and refined (Patil et al., 2023). However, as technology has evolved, new tools have emerged to augment, accelerate, and transform this established approach, particularly through artificial intelligence. The recent integration of AI-assisted tools in filmmaking offers novel efficiencies and creative possibilities, streamlining workflows without undermining the traditional creative process.

In scriptwriting, for example, traditional methods involve brainstorming, outlining, drafting, and revising by the writer or writing team, a process that can be time-consuming and reliant on human creativity for dialogue and plot development. Today, AI tools like ChatGPT and Sudowrite assist by generating plot ideas, refining dialogue, and even suggesting scene structure based on prompts. ChatGPT can quickly provide character backgrounds, suggest story arcs, or offer alternative plot directions, helping writers accelerate early stages of brainstorming and ideation. Sudowrite goes a step further, offering a collaborative experience that allows writers to enhance character development and dialogue while retaining creative control over the narrative. These tools do not replace the vision of the writer but enhance it, enabling writers to iterate and explore possibilities more efficiently (Dayo et al., 2023).

Pre-production traditionally involves detailed visualization, from concept art to location scouting and set design, all aimed at solidifying the visual style and tone of the film. Historically, this requires artists and designers to create initial sketches, paintings, or models, which the director then approves or revises. AI now supports this phase with tools like Runway and Stable Diffusion, which generate concept art, character designs, and environmental visuals based on textual prompts. Runway allows directors and designers to experiment with various visual styles, settings, and atmospheres, enabling a quick and flexible approach to world-building. Stable Diffusion similarly helps filmmakers generate high-quality images that capture potential set layouts, costumes, and character appearances, providing a cohesive aesthetic vision early in the creative process. These AI-driven tools allow the production team to visualize concepts and share ideas more effectively, reducing the time and resources required for initial design (Kim et al., 2024).

Storyboarding and shot composition, traditionally labor-intensive processes, rely on detailed sketches and visual references to plan camera angles, framing, and scene flow. Traditionally, storyboard artists work closely with directors to create these visual representations. Today, tools like Storyboard That and ShotDeck assist by automating aspects of this process. Storyboard That offers an intuitive, AI-enhanced interface for building digital storyboards that match the visual intent of the director, while ShotDeck uses AI to recommend visual references based on the scene's mood or style. These tools enable directors to experiment with shot layouts and framing options early in the production, ensuring a well-structured visual plan that guides the filming process. Through streamlining the storyboarding phase, such tools allow directors to explore visual possibilities and make adjustments in a more efficient manner (Lee, 2023).

Principal photography, or the actual filming process, traditionally depends heavily on on-set decisions about lighting, framing, and shot selection, with limited scope for real-time feedback on audience impact. Tools like Cinelytic and DeepDream now enhance this stage by integrating AI-driven analytics and stylistic effects. Cinelytic provides predictive analytics, helping filmmakers assess which elements may resonate most with audiences, while DeepDream applies filters and stylistic effects directly to footage, allowing for real-time experimentation with visual styles. These tools offer immediate feedback and creative flexibility, giving filmmakers insights that would typically require post-production adjustments (Lama, 2024).

In the editing and post-production phase, traditional techniques require extensive human involvement for tasks like color correction, audio balancing, and visual effects integration. AI-assisted tools, including Adobe Premiere Pro's Sensei and DaVinci Resolve, are now reshaping these tasks. Adobe Sensei automates tasks like color grading and scene transitions, freeing editors to focus on creative refinement rather than technical corrections. Similarly, DaVinci Resolve offers AI-driven features for color grading and noise reduction, enhancing the visual quality of footage. These AI capabilities allow editors to produce polished, cohesive sequences with greater efficiency, thereby shortening the post-production timeline and enabling a sharper focus on narrative cohesion and emotional impact (Wu et al., 2023).

For VFX and CGI, traditional processes demanded highly specialized skills and considerable time to create complex visual elements. Now, tools such as DeepArt and GANPaint Studio streamline VFX production by applying intricate styles to footage or altering elements within scenes. DeepArt can transform live-action footage into surreal, painterly visuals, while GANPaint Studio enables artists to seamlessly add or remove objects, like trees or buildings, from a scene with a high degree of realism. These tools expand the creative possibilities within VFX, allowing artists to design detailed environments without the resource-intensive demands of traditional CGI (Yi et al., 2024).

Finally, sound design, an essential but often labor-intensive component of post-production, has also been transformed by AI. Traditional sound editing requires careful synchronization, mixing, and often original score composition. Tools like AIVA and Descript support these tasks by automating aspects of sound creation and editing. AIVA can generate original musical compositions tailored to the film's emotional landscape, while Descript facilitates precise audio editing, including voice cloning and transcription. These tools reduce the time and resources needed for sound design, allowing filmmakers to achieve high-quality audio that enhances the narrative and emotional resonance of the film (Löbbers et al., 2023).

In distribution and audience analysis, traditional methods involve market research and focus groups to anticipate viewer response. AI-powered platforms like Grokstyle and IBM Watson now analyze visual preferences and predict audience demographics, helping studios make informed decisions about release strategies. Grokstyle's visual search technology identifies which visual styles may appeal to specific demographics, while IBM Watson provides predictive insights into audience preferences, ensuring that the film reaches its target market effectively. By leveraging AI insights, filmmakers can better align their distribution approach with audience expectations, maximizing the reach and impact of the film (Sookhom et al., 2023). Together, these AI-assisted tools augment the traditional filmmaking process, enabling greater efficiency, creativity, and adaptability across each production phase. By blending established practices with advanced technology, filmmakers gain access to a robust toolkit that enhances storytelling potential while preserving the artistic integrity of traditional cinematic methods.

The integration of AI tools into the creative workflow offers filmmakers and video artists unprecedented opportunities to streamline production, enhance creativity, and navigate intellectual property concerns. Traditionally, the film production process involves distinct stages, each contributing to the overarching narrative and visual storytelling. However, the advent of GAI has introduced new workflows, where creators can utilize tools like MidJourney, Stable Diffusion, and Runway to augment their traditional processes in unique ways, from concept design to final edits (Wang et al., 2024). Through an understanding and documenting each phase of the AI-enhanced workflow, creators can both ensure the copyrightability of their work and clarify their authorship. This approach aligns with emerging legal frameworks that recognize human intervention as essential for establishing originality and ownership in AI-assisted creations.

Considered together, creatives can navigate the processes and potential for these new tools (Table 2.3). In traditional filmmaking, the initial stages begin with concept development and pre-production planning. For video artists and filmmakers using AI, this process can now begin with text-to-image generators like MidJourney or Stable Diffusion, which allow creators to visualize early concepts from textual prompts. Artists craft these prompts to specify artistic style, composition, and thematic details, providing the AI with a conceptual framework. The prompt design process, when detailed and deliberate, helps establish the originality of the human creator, making the work eligible for copyright protection as it reflects the artistic vision of the creative. In fact, in prior copyright cases, only the prompts to create an image were copyrightable as they demonstrated the "creative intent" of the author. This phase also mirrors the traditional process of concept art creation but introduces greater flexibility and speed in visualizing multiple ideas.

Once an image has been generated, the next step involves integrating these visuals into a coherent video sequence. AI platforms like Runway offer image-to-video capabilities, which allow artists to animate static images created by MidJourney or Stable Diffusion, translating still visuals into dynamic motion.

Table 2.3 AI-enhanced filmmaking workflow.

Stage	Traditional Process	AI-Enhanced Tools	Human Contribution
Concept Development	Developing story ideas, visual themes, and artistic direction.	**MidJourney, Stable Diffusion**	Crafting detailed prompts that define style, composition, and thematic direction, showcasing original vision.
Pre-Production	Creating concept art, storyboarding, and planning scenes.	**Text-to-Image Generators**	Selecting specific generated images to align with artistic intent and narrative flow.
Image Animation	Traditionally done with drawn or computer-generated animations.	**Runway**	Applying camera movements, transitions, and lip-sync to bring images to life, enhancing narrative continuity.
Scene Assembly	Arranging video clips, applying transitions, and color grading.	**Adobe Premiere Pro, DaVinci Resolve**	Editing AI-generated clips for cohesion, refining pacing, and ensuring continuity in storytelling.
Sound Design & Scoring	Integrating music, sound effects, and voiceovers to convey emotional depth and narrative tone.	**AIVA** (Music), **Descript** (Dialogue)	Choosing and editing sound to match visuals, controlling timing, tone, and atmosphere.
Final Editing & Refinement	Finalizing all elements for smooth storytelling, including final transitions and color adjustments.	**Premiere Pro, DaVinci Resolve**	Ensuring cohesive storytelling and aesthetic harmony across all elements for a polished final product.
Documentation	Recording each stage of the creative process for legal and organizational purposes.	**Manual documentation**	Detailing each step and decision made to highlight human input, supporting copyright eligibility.

Runway's tools for advanced camera control, lip-syncing, and scene transitions enable video creators to add depth and temporal coherence to their narratives, allowing AI-generated scenes to interact and evolve. For example, Runway's camera control tools can simulate complex camera movements, enhancing the cinematic feel and aligning with traditional techniques used by directors and cinematographers (Ma, 2024). This aspect of the process requires human oversight and creative choices regarding pacing, angle, and sequencing, underscoring the artist's role in shaping the final product.

Following the generation of animated sequences, the clips are then exported into post-production software, such as Adobe Premiere Pro or DaVinci Resolve. Here, filmmakers focus on refining continuity, ensuring visual consistency, and enhancing storytelling impact. This editing phase mirrors traditional workflows, where continuity editing, scene transitions, and pacing are meticulously crafted to align with the narrative arc. AI-generated content is assembled in Premiere Pro to maintain cohesion, and editors apply color grading, effects, and scene transitions to achieve a seamless viewing experience. At this stage, human intervention is crucial in determining the arrangement and impact of scenes, ensuring that the video remains aligned with the creator's vision and intent. This hands-on engagement with the AI-generated content solidifies the human element in the work, bolstering claims for copyright by documenting the creative input of the artist (Erpelding et al., 2024).

In the final phase, sound design and dialogue are layered over the visual content. Tools like AIVA generate background music tailored to the mood and tempo of the scenes, while software such as Descript assists with precise audio editing, lip-syncing, and even voice cloning if required. These sound elements enhance the emotional and narrative depth of the video, a critical aspect of cinematic storytelling. Through selecting, editing, and integrating sound in alignment with the visual sequence, the filmmaker further reinforces their authorship, showcasing their aesthetic choices and artistic vision. This sound design process mirrors traditional workflows in film and emphasizes the significance of human input, as even AI-generated soundscapes are guided by the thematic and emotional objectives of the creator (Lac and Moura, 2024).

Throughout the AI-augmented creative workflow, meticulous documentation is essential to ensure that each stage reflects the creator's role and decisions. Filmmakers should document prompt designs, capture their choices in selecting specific outputs, and record any alterations or refinements made in post-production. This record of creative steps—from generating initial visuals to integrating animation, editing, and sound design—serves as tangible proof of the human contribution, reinforcing the copyright eligibility of the AI-assisted work. Through a combination of traditional cinematic techniques with advanced AI tools, creators can bridge the gap between human artistry and algorithmic generation. This approach preserves the essence of cinematic storytelling while embracing the efficiencies and possibilities of generative technologies, ensuring that AI serves as an augmentation rather than a replacement for human creativity. The resulting works not only align with current copyright standards but also establish a framework for integrating AI into future creative practices, balancing innovation with the enduring values of human originality and artistic vision.

Moreover, the intricate dance between creativity and technology within generative tools marks a significant evolution in the filmmaking process, presenting both a powerful opportunity and a set of challenges for artists and directors alike. The ability of GAI to craft dynamic video content from textual input has opened up a realm of possibilities, much like the art generators did for still images. While these advancements appear to have emerged suddenly, they represent the latest step in a longstanding tradition of innovation within the cinematic arts, where each new technological leap redefines the boundaries of storytelling. Central to this transformation is the delicate balance of human input and AI assistance. The considerations surrounding authorship, originality, and ethical engagement with AI serve as crucial elements for contemporary filmmakers navigating this new landscape. By documenting their specific contributions, from prompt design to creative curation and post-production refinement, artists and filmmakers not only maintain control over the creative process but also ensure their roles remain vital in the aesthetic and narrative impact of the output. This approach to AI-assisted filmmaking empowers creatives to capture and direct these new capabilities while retaining the authenticity of human vision and artistic judgment.

This confluence of human creativity and machine precision brings unique advantages, yet it also necessitates a thoughtful approach to ethical and legal concerns, including issues of copyright and cultural representation. By understanding these elements within the broader historical context of cinematic technological progression, filmmakers gain insights into how new generative tools align with or diverge from traditional film practices. As these technologies push the boundaries of what is possible on screen, the next chapter will look into the historical foundations that frame this drive for innovation, illustrating how each advancement has built upon previous breakthroughs and paved the way for AI to become the latest transformative force in visual storytelling.

References

Aaland, M. (2006). *Photoshop Elements 3 Solutions: The Art of Digital Photography*. John Wiley & Sons.

Abbott, R. and Rothman, E. (2023). Disrupting creativity: Copyright law in the age of generative artificial intelligence. *Florida Law Review,* 75: 1141.

Abdikhakimov, I. (2023, June). Unraveling the Copyright conundrum: Exploring AI-generated content and its implications for intellectual property rights. pp. 18–32. *In*: *International Conference on Legal Sciences* (Vol. 1, No. 5).

Ahmed, D. (2022). Senses, experiences, emotions, memories: artificial intelligence as a design instead of for a design in contemporary Japan. *Intelligent Buildings International,* 14(2): 133–150.

Ajani, G. (2022, May). Human authorship and art created by artificial intelligence–where do we stand? pp. 253–270. *In*: *Digital Ethics*. Nomos Verlagsgesellschaft mbH & Co. KG.

Akkus, C., Chu, L., Djakovic, V., Jauch-Walser, S., Koch, P., Loss, G. et al. (2023). Multimodal deep learning. *arXiv preprint arXiv:2301.04856.*

Amankwah-Amoah, J., Abdalla, S., Mogaji, E., Elbanna, A. and Dwivedi, Y. K. (2024). The impending disruption of creative industries by generative AI: Opportunities, challenges, and research agenda. *International Journal of Information Management,* 102759.

Ansari, T. (2022). How AI transformed the art world in 2022. *Analytics India Magazine (AIM)* October 30. Retrieved from: https://analyticsindiamag.com/how-ai-transformed-the-art-world-in-2022/.

Ananthanagu, U. and Agarwal, P. (2023). A systematic review and future perspective of mental illness detection using artificial intelligence on multimodal digital media. *Intelligent Sustainable Systems: Selected Papers of WorldS4 2022*, 1: 35–46.

Boden, M. A. (2010). *Creativity and Art: Three Roads to Surprise*. Oxford University Press.

Brittain, B. (2024). Artist sues after US rejects copyright for AI-generated image. *Reuters,* September 26, 2024: https://www.reuters.com/legal/litigation/artist-sues-after-us-rejects-copyright-ai-generated-image-2024-09-26/.

Carnovalini, F. and Rodà, A. (2020). Computational creativity and music generation systems: An introduction to the state of the art. *Frontiers in Artificial Intelligence,* 3: 14.

Cheng, M. (2022, April). The creativity of artificial intelligence in art. *In*: *Proceedings* (Vol. 81, No. 1, p. 110). MDPI.

Csikszentmihalyi, M. and Csikszentmihalyi, M. (2014). *Society, Culture, and Person: A Systems View of Creativity* (pp. 47–61). Springer Netherlands.

Collet-Sabé, J. (2023). Pre-modern epistemes inspiring a new Global Sociology of Education Imagination. *British Journal of Sociology of Education,* 44(8): 1249–1266.

Cope, D. (2015). Computational creativity and music. *Computational Creativity Research: Towards Creative Machines,* 309–326.

Dayo, F., Memon, A. A. and Dharejo, N. (2023). Scriptwriting in the age of AI: Revolutionizing storytelling with artificial intelligence. *Journal of Media & Communication,* 4(1): 24–38.

Di Mitri, D., Schneider, J. and Drachsler, H. (2022). The rise of multimodal tutors in education: insights from recent research. *Handbook of Open, Distance and Digital Education,* 1–20.

Du Sautoy, M. (2019). The creativity code: Art and innovation in the age of AI. *In*: *The Creativity Code.* Harvard University Press.

Duch, W. (2006, July). Computational creativity. pp. 435–442. *In*: *The 2006 IEEE International Joint Conference on Neural Network Proceedings*. IEEE.

Dwivedi, P. S. (ed.). (2021). *Aesthetics and the Philosophy of Art: Comparative Perspectives*. Taylor & Francis.

Epstein, Z., Hertzmann, A., Investigators of Human Creativity, Akten, M., Farid, H., Fjeld, J. and Smith, A. (2023). Art and the science of generative AI. *Science,* 380(6650): 1110–1111.

Erpelding, C., Beck, J., Swerzenski, J. D. and Brecheisen, T. (2024). Forum on artificial intelligence. *Journal of Film and Video,* 76(1): 44–55.

Escalante-De Mattei, S. (2024). Stephen Thaler's quest to get his 'autonomous' AI legally recognized could upend copyright law forever. *Art in America*. January 8, 2024: https://www.artnews.com/art-in-america/features/stephen-thaler-quest-ai-legally-recognized-upend-copyright-law-1234692243/.

Feuerriegel, S., Hartmann, J., Janiesch, C. and Zschech, P. (2024). Generative AI. *Business & Information Systems Engineering,* 66(1): 111–126.

Foucault, M. (1969). 1979. What is an author. *Textual Strategies,* 141–60.

Gordon, C. (2023) ChatGPT Is The Fastest Growing App in The History of Web Applications. *Forbes,* February 2, 2023: https://www.forbes.com/sites/cindygordon/2023/02/02/chatgpt-is-the-fastest-growing-ap-in-the-history-of-web-applications/.

Harrison, C. (2003, January). The evolving digital workflow in cinema. pp. 313–318. *In*: Color and Imaging Conference (Vol. 11). *Society of Imaging Science and Technology*.

Hazucha, B. (2022). Artificial intelligence and cultural production: possible impacts on creativity and copyright law. *Available at SSRN 4028106.*

Hong, J. W. and Curran, N. M. (2019). Artificial intelligence, artists, and art: attitudes toward artwork produced by humans vs. artificial intelligence. *ACM Transactions on Multimedia Computing, Communications, and Applications (TOMM),* 15(2s): 1–16.

Hutson, J., Lively, J., Robertson, B., Cotroneo, P. and Lang, M. (2023). Of Techne and Praxis: Redefining Creativity. pp. 21–36. *In*: *Creative Convergence: The AI Renaissance in Art and Design*. Cham: Springer Nature Switzerland.

Jennings, K. E. (2010). Developing creativity: Artificial barriers in artificial intelligence. *Minds and Machines,* 20: 489–501.

Jordanous, A. (2016). Four PPPPerspectives on computational creativity in theory and in practice. *Connection Science,* 28(2): 194–216.

Kelby, S. (2020). *The Digital Photography Book* (Vol. 1). Rocky Nook, Inc.

Kharchenko, P., Chibalashvili, A., Savchuk, I., Sydorenko, V. and Khasanova, I. (2023). Technologies as a mediator between creator and audience in postmodern art practices. BRAIN. *Broad Research in Artificial Intelligence and Neuroscience,* 14(1): 500–514.

Kim, H., Ali, G., Han, B., Kim, H. Y., Kim, J., Shin, H. et al. (2024). ASAP for multi-outputs: auto-generating storyboard and pre-visualization with virtual actors based on screenplay. *Multimedia Tools and Applications,* 1–24.

Kurt, D. E. (2018). *Artistic Creativity in Artificial Intelligence*. Master diss., Radbound University.

Lac, V. and Moura, F. T. (2024). Sentiment Analysis of AI Generated Music Using Latent Dirichlet Allocation (LDA). *AIMC 2024* (09/09–11/09).

Lama, R. (2024). The Future of Virtual Backgrounds in Filmmaking.

Lazer, D., Hargittai, E., Freelon, D., Gonzalez-Bailon, S., Munger, K., Ognyanova, K. et al. (2021). Meaningful measures of human society in the twenty-first century. *Nature,* 595(7866): 189–196.

Lee, S. (2023). Enhancing video storyboarding with artificial intelligence: an integrated approach using ChatGPT and Midjourney within AiSAC. *International Journal of Advanced Culture Technology,* 11(3): 253–259.

Lehman, R. S. (2020). Criticism and judgment. *ELH,* 87(4): 1105–1132.

Lescrauwaet, L., Wagner, H., Yoon, C. and Shukla, S. (2022). Adaptive legal frameworks and economic dynamics in emerging technologies: Navigating the intersection for responsible innovation. *Law and Economics,* 16(3): 202–220.

Löbbers, S., Barthet, M. and Fazekas, G. (2023). AI as mediator between composers, sound designers, and creative media producers. *arXiv preprint arXiv:2303.01457.*

Ma, Y. (2024). *Image and Video Generative AI in Filmmaking* (Master's thesis, New York University Tandon School of Engineering).

Makam, G. and Dutta, R. (2023). AI-generated creations: navigating legal implications and crafting effective policy frameworks. *Available at SSRN 4520938.*

Mazzone, M. and Elgammal, A. (2019, February). Art, creativity, and the potential of artificial intelligence. *In*: *Arts* (Vol. 8, No. 1, p. 26). MDPI.

Moreno, H., Gómez, A., Altares-López, S., Ribeiro, A. and Andújar, D. (2023). Analysis of stable diffusion-derived fake weeds performance for training Convolutional Neural Networks. *Computers and Electronics in Agriculture,* 214: 108324.

Mori, M., MacDorman, K. F. and Kageki, N. (2012). The uncanny valley [from the field]. *IEEE Robotics & Automation Magazine,* 19(2): 98–100.

Murphy, B. (2022). Is Lensa AI stealing from human art? An expert explains the controversy. *Science Alert.* 15 December 2022. Retrieved: https://www-sciencealert-com.cdn.ampproject.org/c/s/www.sciencealert.com/is-lensa-ai-stealing-from-human-art-an-expert-explains-the-controversy/amp.

Orr, J. (2020). The discarded mind: from divine ideas to secular concepts. *Neue Zeitschrift für Systematische Theologie und Religionsphilosophie,* 62(2): 167–189.

Patil, B. D., Patil, S. P. and Suryawanshi, R. R. (2023). Exploring the visual art of filmmaking. *Journal of Survey in Fisheries Sciences,* 10(3): 698–703.

Pinel, F. and Varshney, L. R. (2014). Computational creativity for culinary recipes. pp. 439–442. *In*: *CHI'14 Extended Abstracts on Human Factors in Computing Systems.*

Ploman, E. W. and Hamilton, L. C. (2024). *Copyright: Intellectual Property in the Information Age.* Taylor & Francis.

Quang, J. (2021). Does training AI violate Copyright law?. *Berkeley Technology Law Journal,* 36*:* 1407.

Ren, J., Xu, H., He, P., Cui, Y., Zeng, S., Zhang, J. et al. (2024). Copyright protection in generative AI: A technical perspective. *arXiv preprint arXiv:2402.02333.*

Risi, S. and Togelius, J. (2020). Increasing generality in machine learning through procedural content generation. *Nature Machine Intelligence,* 2(8): 428–436.

Rykiel Jr, E. J. (1989). Artificial intelligence and expert systems in ecology and natural resource management. *Ecological Modelling,* 46(1-2): 3–8.

Säljö, R. (2010). Digital tools and challenges to institutional traditions of learning: technologies, social memory and the performative nature of learning. *Journal of Computer Assisted Learning,* 26(1): 53–64.

Shahbazi, N., Lin, Y., Asudeh, A. and Jagadish, H. V. (2023). Representation bias in data: A survey on identification and resolution techniques. *ACM Computing Surveys,* 55(13s): 1–39.

Sookhom, A., Klinthai, P., A-masiri, P. and Kerdvibulvech, C. (2023). A new study of AI artists for changing the Movie Industries. *Digital Society,* 2(3): 37.

Spica, E. (2024). Public interest, the true soul: Copyright's fair use doctrine and the use of Copyrighted works to train generative AI tools. *Texas Intellectual Property Law Journal,* 33(1).

Sturm, B. L., Iglesias, M., Ben-Tal, O., Miron, M. and Gómez, E. (2019, September). Artificial intelligence and music: open questions of copyright law and engineering praxis. *In*: *Arts* (Vol. 8, No. 3, p. 115). MDPI.

Van Dijck, J. (2005). From shoebox to performative agent: the computer as personal memory machine. *New Media & Society,* 7(3): 311–332.

Vincent, J. (2023). Getty Images is suing the creators of AI art tool Stable Diffusion for scraping its content. *The Verge.* January, 17 2023: https://www.theverge.com/2023/1/17/23558516/ai-art-copyright-stable-diffusion-getty-images-lawsuit.

Wang, H., Smith, D. and Kudelska, M. (2024, August). 10x future of filmmaking empowered by AIGC. *In*: *2024 IEEE 7th International Conference on Multimedia Information Processing and Retrieval (MIPR)* (pp. 68–74). IEEE.

Wu, J., Gan, W., Chen, Z., Wan, S. and Lin, H. (2023). AI-generated content (aigc): A survey. *arXiv preprint arXiv:2304.06632.*

Yi, C. Y. and Isa, A. N. M. (2024). Mapping the landscape of VFX and AI: A PRISMA-guided systematic review. *International Journal of Business and Technology Management,* 6(3): 114–132.

Znidarsic, M., Cardoso, A., Gervás, P., Martins, P., Hervás, R., Alves, A. O. et al. (2016, June). Computational creativity infrastructure for online software composition: A conceptual blending use case. pp. 371–379. *In*: *International Conference on Computational Creativity*. Sony CSL Paris.

Chapter 3
Historical Context and Theoretical Perspectives

This chapter traces the evolution of AI in video creation, revealing how the generative technologies of today continue a long history of automated artistry. From ancient automata like the Draughtsman-Writer to twentieth-century pioneers in algorithmic art such as Harold Cohen and Nam June Paik, each era reflects a fascination with machines that could create independently and/or augment human creatives. The advent of digital tools in the 1960s enabled early explorations in computer-generated art, paving the way for AI-driven techniques that now redefine visual storytelling in cinema and interactive media. Advances in CGI, machine learning, and immersive technologies have integrated AI and GAI into the filmmaking process, from dynamic effects to narrative construction. By viewing contemporary developments in generative visuals through this historical lens, the chapter situates modern tools as extensions of the human pursuit of creativity through technology, setting the stage for future discussions on the integration of human and machine creativity in video production.

3.1 Ancient Origins of Creative Machines

The influence of generative AI technologies on video creation is growing rapidly, transforming not only the technical aspects of production but also reshaping the very fabric of storytelling and audience engagement. Seemingly emerging out of nowhere in 2023, these video platforms have evolved from basic support tools to sophisticated creative partners, expanding beyond initial applications in editing and effects to encompass comprehensive roles throughout production. Today, these systems contribute to every phase of the video production process, from pre-visualization and concept development to real-time editing and post-production enhancements. The release of the Gen-1 model from Runway in early 2023 marked a significant milestone in this evolution. Known for its contributions to developing Stable Diffusion, Runway introduced Gen-1 as a pioneering tool that could transform existing video clips by applying various styles through text prompts or reference images. Alongside platforms like Pictory, DALL-E, Midjourney, and Pika, Gen-1 enables creators to turn simple text prompts into high-quality images and then into cinematic visuals, allowing for the production of compelling content with minimal resources (Gao and Liu, 2023; Huang et al., 2023; Reddy et al., 2024).

The accessibility of these advanced tools has expanded the pool of storytellers, giving rise to a new generation of independent creators who now have the capability to produce professional-grade videos. Through the automation of certain complex production tasks, AI-assisted systems empower creators, augmenting their abilities should they lack traditional training or large production teams to produce impactful content. Generative tools, for instance, offer real-time scene transitions, CGI rendering, and even actor simulation, which were once confined to high-budget studios. As a result, individuals and smaller teams can now compete in a market that was previously dominated by larger production houses, and video artists can visualize innovative content like never before. The shift not only accelerates the production process but also lowers costs, fostering innovation across various genres from marketing to independent filmmaking (McAlister et al., 2024; Sager, 2024; Tao et al., 2024). In essence, this integration into video creation is opening doors for more diverse narratives and creative expressions, signaling a major shift in the video production landscape.

At the same time, it is important to note that artists have long had a complex relationship with machines, technology, and now AI, serving as both pioneers and critics of technology-driven art forms. Today, many artists express concerns about AI art generators, questioning the authenticity, authorship, and ethical implications of using automated tools to replicate their unique styles. Some fear that the proliferation of AI-generated art undermines the human touch, creativity, and individuality that traditionally define artistic expression. However, this tension is not new; artists have historically embraced emerging technologies, from photography to digital media, often pushing these innovations in ways that expand artistic boundaries. In fact, artists were among the first to experiment with algorithmic and generative art, with pioneers like British-born Harold Cohen (1928–2016), who developed the AI program AARON in the 1970s to create abstract art autonomously (Cohen, 1995; Mazzone and Elgammal, 2021). Over several decades, Cohen continually refined AARON, enhancing its capabilities from producing simple abstract forms to creating complex, representational images. This evolution was marked by significant milestones, including the ability of the program to autonomously apply color and develop compositional rules. Such early explorations laid the groundwork for modern models such as DALL-E, Midjourney, and Stable Diffusion, which now enable the generation of intricate visual compositions from textual descriptions. But the history of creative machines can be traced back well beyond the last century.

The conceptual origins of artificial intelligence in art trace back to early myths and mechanical inventions that blurred the boundaries between human artistry and programmed design. The roots of automated creativity, while commonly associated with modern advancements, therefore, extend deeply into antiquity through the efforts of early visionaries who conceptualized machines capable of simulating human actions. Figures such as Daedalus and Hero of Alexandria exemplify these ancient innovators. Daedalus (first mentioned around 1400 BCE), renowned in

Greek mythology as a skillful architect and craftsman, is credited with crafting life-like statues and self-moving creations. These statues moved their limbs as though alive, driven by hidden mechanisms that imitated human motion, inspiring a legacy that questioned the boundaries between animate and inanimate (Gerolemou, 2022). He also designed intricate labyrinths, including the famed labyrinth of Crete for King Minos, which was so complex it was nearly impossible to navigate, even for its creator, and housed the half-man half-bull Minotaur (Fitzgerald, 1984). After he and his son Icarus were imprisoned in the labyrinth by the kind, Daedalus was credited with inventing artificial wings, crafted from feathers and wax, that allowed humans to soar through the air—an early attempt at mechanized flight that later influenced countless myths and stories. However, the invention also had tragic consequences, as seen in the death of his son Icarus, who flew too close to the sun with the artificial wings and fell, thus illustrating the duality of technology as both a source of marvel and a potential peril (Chiglintsev, 2015).

The Hellenistic period was particularly significant for mechanical artistry, with inventors employing engineering techniques to create awe-inspiring automata. Hero (Heron) of Alexandria (1st-2nd century CE), the Greek mathematician and engineer, pioneered mechanical automation, developing devices such as the aeolipile, a primitive steam engine, and automated theater displays, which featured lifelike performances through intricate mechanical designs (Moussa and Fekry, 2022). Other entertaining mechanisms such as the automatic water organ and coin-operated machines, designed to awe and inspire belief in the supernatural capabilities of mechanics. The automata of the engineer (Fig. 3.1), which involved elaborate systems of weights, counterweights, and hydraulic technology, represented an early understanding of machine operation that foreshadowed principles used in later robotics and computing. The significance of these creations lies not only in their technical achievements but also in their influence on philosophical and ethical discussions, particularly concerning human interaction with automated entities and the nature of creative autonomy (McCourt, 2012).

These devices, along with mythological figures like Talos—a bronze giant said to be crafted by the forge-god Hephaestus to guard Crete—reflect an ancient cultural fascination with mechanized life (Vasileiadou et al., 2003). Such figures symbolized protection and service but also evoked anxieties about control and autonomy. Ancient legends, like the myth of Pandora also crafted by Hephaestus, and Archytas' flying pigeon, an early self-propelled device, demonstrate a deep-rooted interest in artificial beings capable of independent function (Moran, 2011). These mythological and mechanical precedents established a cultural framework that would ultimately support the emergence of AI in creative domains, reflecting an enduring belief that machines might replicate or even enhance human creativity. This legacy illustrates the allure and ambiguity of machine creativity, as ancient innovators sought to push the boundaries of human expression by delegating it to programmable entities.

Throughout the Middle Ages and renaissance, cultures in both Western and non-Western worlds created intricate automata that merged artistry with technological innovation. In the Islamic world, engineers like Al-Jazari (1136–1206) and the Banu Musa brothers pioneered advanced mechanical devices, many of

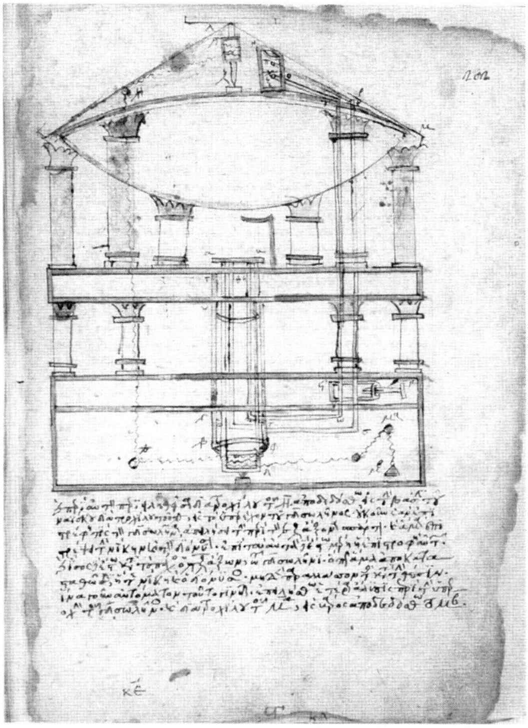

Figure 3.1 Hero of Alexandria, Automata, Biblioteca Marciana, Gr. 516, fol. 202r.,Venice, early 14th century (Creative Commons License Zero).

which were programmable by early standards. In his 1206 *Book of Knowledge of Ingenious Mechanical Devices,* Al-Jazari describes over 50 inventions, including a water-powered orchestra where musicians played music on a boat to entertain guests, showcasing a complex system of rotating drums with pegs to control movements. His famed Elephant Clock (Fig. 3.2), a water clock adorned with mechanical figures, exemplified the blend of artistry and engineering of the era (Uzun and Vatansever, 2008). The three Banu Musa brothers, operating in 9th-century Baghdad, created an automatic flute player powered by steam, another early example of programmable mechanics (Nordin and Ramli, 2020). These devices highlighted the contributions to automata by the Islamic world, combining functional innovation with aesthetic appeal.

The Byzantine Empire also embraced automata to display imperial power and sophistication. The imperial palace of Emperor Theophilos (ca. 812–842) in 9th-century Constantinople famously featured mechanical lions that roared and birds that chirped from a gilded bronze tree, creating an awe-inspiring spectacle for visitors (Filson, 2017). Similarly, the Byzantine Throne of Solomon, as described by Liudprand of Cremona (ca. 920–972) in 949 AD, incorporated mechanical lions and birds around the imperial throne, symbolizing royal authority and divine favor. These displays reflected the Byzantine fascination with technological wonders that reinforced imperial grandeur, as automata became political symbols as much

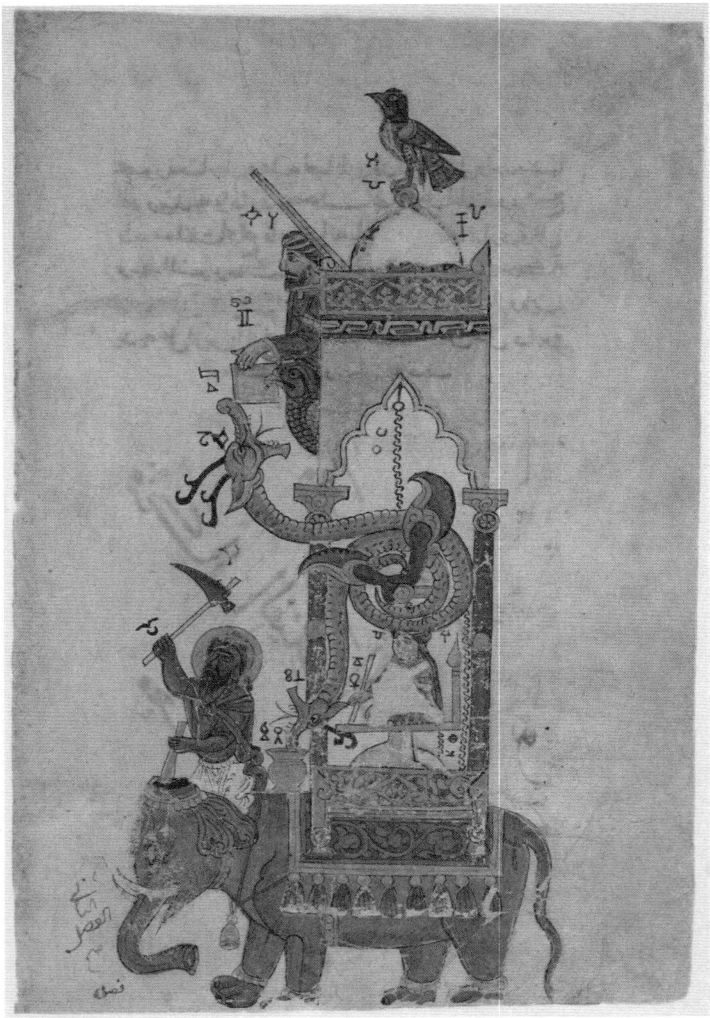

Figure 3.2 Al-Jazari, Elephant Clock, *Book of Knowledge of Ingenious Mechanical Devices*, 1315 (Creative Commons License Zero).

as engineering feats (Iafrate, 2016). The Byzantine court used these creations to showcase their technological prowess and impress foreign dignitaries.

In renaissance Europe, the revival of classical knowledge and an expanding interest in mechanical engineering inspired new approaches to automata. The designs of Leonardo da Vinci (1452–1519) from 1495, such as his Mechanical Knight capable of standing, sitting, and moving its arms, illustrate another sophisticated exploration of mechanics (Ren, 2023). Giovanni Fontana (ca. 1395–1455), an Italian engineer, created fantastical devices such as a puppet driven by a clothed primate, blending art and technology (Buchan, 2013). Central European clockmakers in cities like Augsburg crafted intricate automata in the form of *nefs* (ornate, ship-shaped table ornaments) and humanoid figures that could play music, reflecting the regional skill in blending utility with decoration (Springmann, 2020). These renaissance automata underscored the fascination with mechanics and artistry, aligning with the broader pursuit of innovation of the period.

These developments in automata were not isolated; instead, they reflect a network of cultural exchange that connected different regions. Automata from the Islamic world reached Europe as diplomatic gifts, such as the water fountain given by Harun al-Rashid (ca. 763–809) to Charlemagne (748–814), which captivated European audiences and inspired local craftsmanship (Greenhalgh, 2008). Byzantine envoys brought stories of Abbasid automata back to Constantinople, fueling curiosity and innovation within their own courts. This exchange helped cultivate a shared fascination with programmable machines and laid the groundwork for future advancements in automata and mechanical arts (Heilo, 2022). Across cultures, these early creations illustrated humanity's drive to simulate life through technology, establishing a legacy of programmable creativity that continues to influence discussions on automation and artificial intelligence today.

During the Age of Enlightenment, advances in mechanical engineering led to increasingly sophisticated automata that simulated human skills, paving the way for future explorations in automated creativity. Henri Maillardet (1745–1830), a Swiss mechanician, long before ChatGPT created an automaton known as the Draughtsman-Writer automaton (Fig. 3.3) in the late 18th century. Now preserved at the Franklin Institute in Philadelphia, this machine could produce four intricate drawings and write three poems, demonstrating a remarkable level of mechanical precision for its time. While it did not create new work, it captivated audiences by mimicking human handwriting and artistry, showcasing the fascination with machines that could imitate human actions (de Panafieu, 1984). This fascination marked a pivotal moment in the journey toward understanding the potential of machines as artistic tools, inspiring further developments in mechanical and computational artistry.

In the 1840s, Ada Lovelace (1815–1852) contributed a visionary perspective that extended beyond mechanical imitation, suggesting that machines could eventually engage in creative processes of their own. Collaborating with Charles Babbage (1791–1871) on his Analytical Engine (Fig. 3.4), Lovelace foresaw the potential for machines to compose music or create visual art, given the right programming. She theorized that machines could generate original outputs, a concept that would later be foundational to artificial intelligence. Although her ideas remained theoretical during her lifetime, Lovelace's insights laid the groundwork for future innovations in AI and computer science by shifting the conversation from replication to creative generation. Her perspective offered a profound philosophical leap, imagining machines as partners in the creative process, not simply tools of reproduction (Aiello, 2016).

By the mid-20th century, inventors explored machines that could respond dynamically to human input, reflecting Lovelacian vision of interactive, adaptive creativity. Gordon Pask (1928–1996), a British cybernetician, developed the MusiColour machine in the 1950s, which used sound input from a human performer to control colored lights, creating a responsive visual display (Pickering, 2007). Around the same time, Jean Tinguely (1925–1991), a Swiss artist, created painting machines

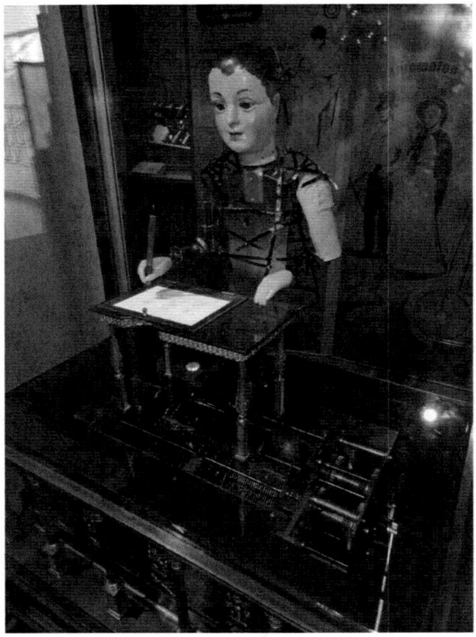

Figure 3.3 Henri Maillardet, *Draughtsman-Writer* (*Automaton*), 1810, London, England; Franklin Institute, Philadelphia, Pennsylvania, USA (Creative Commons License Zero).

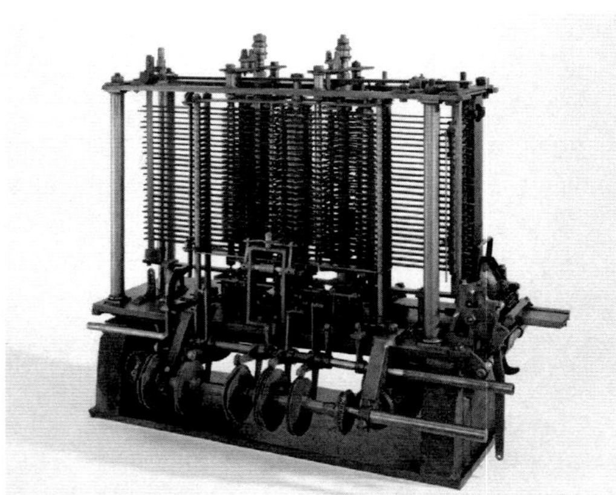

Figure 3.4 Charles Babbage, Analytical Engine, 1834–1871 (Creative Commons License 2.0).

that allowed users to select variables such as color, position, and duration, while the machine generated abstract artworks autonomously. These devices produced unique and unpredictable pieces, blurring the line between human and machine creativity (Voland, 1990). Together, these innovations bridged the gap between mechanized imitation and autonomous creative processes, building upon the foundational concepts introduced by figures like Lovelace and Maillardet, and setting the stage for the establishment of artificial intelligence as a field dedicated to replicating—and enhancing—human creativity.

The pre-history of AI art reveals a nuanced relationship with automation, reflecting the persistent tension between the drive for innovation and deeper philosophical questions of authenticity and control. The early examples of programmed creativity—from Maillardet's Draughtsman-Writer to Lovelace's theoretical musings and the reactive machines of Pask and Tinguely—demonstrate a consistent fascination with machines that could mimic or even extend human creative capacities. These inventions provided more than mere technical achievements; they introduced a framework for questioning the nature of creativity itself, prompting reflections on what distinguishes human artistry from machine-generated work. As these early automata and programmable devices paved the way for modern artificial intelligence, they also foreshadowed the ethical and philosophical debates that continue to surround their role in creative fields. The legacy of these pioneering efforts underscores a long-standing human aspiration to push the boundaries of creativity through technology, while simultaneously acknowledging the inherent complexities of sharing artistic agency with machines. These historical precedents thus form a foundational layer for contemporary discussions on AI art, encapsulating both the potential for technological progress and the enduring inquiry into the role, purpose, and boundaries of machine creativity.

3.2 Origins of Video Art (1960s–1970s)

The 1950s marked a pivotal moment in the technological advancements that would ultimately fuel the development of video art, digital art, and interactive media. This era saw the establishment of artificial intelligence as an academic discipline at a 1956 research workshop at Dartmouth College, where early concepts of machine learning and automated processes began to take shape (Krauss, 2024). Around the same time, innovations in computer science and emerging fields like cybernetics laid the groundwork for the integration of computational methods into creative practices. Artists and engineers alike explored the potential of algorithmic and computer-generated art, which was then considered an experimental approach to creativity (Johnston, 2008). This fusion of art and technology inspired a new generation of creators eager to break away from traditional methods, setting the stage for groundbreaking applications of AI in art installations and software. As digital and video art emerged, pioneering artists would dedicate themselves to blending technology with artistic vision, redefining the boundaries of artistic expression and establishing the foundations for what would later become known as video art.

The rise of computer-generated imagery (CGI) and digital art in the 1960s further expanded the realm of artistic possibilities, as artists began to experiment with machines capable of generating visual content. This period marked a shift from using technology merely as a tool to employing it as a collaborator in the creative process. As computer graphics evolved, early practitioners of digital art used programming and algorithms to create abstract, mathematical forms, which were often referred to as "algorithmic art" or "computer art." These creations were among the first instances of art that relied on computational processes to produce original, non-replicative works, challenging traditional concepts of authorship and artistic control (Fuchs and Wenz, 2022). As noted, British artist Harold Cohen, one of the early pioneers in this field, would devote much of his life to developing AARON, an AI-driven art-making program, in his quest to explore the artistic potential of machines (Cohen, 2016).

In parallel with digital and computer art, video art emerged as a radical medium for artists in the 1960s and 1970s who sought to move beyond the constraints of traditional film. Video art offered artists an experimental platform where they could explore new forms of visual storytelling and challenge conventional narratives associated with cinema (Meigh-Andrews, 2013). With its accessible, portable technology, video allowed artists to record, edit, and display moving images without relying on the commercial frameworks of film production. Artists such as Nam June Paik (1932–2006), known as the "father of video art," and others pioneered this medium, using video installations and television sets to craft immersive experiences that blended art and technology. Through video, these artists could engage viewers in real-time and create interactive, often improvisational pieces, pushing the limits of what visual art could convey. The influence of video art in this period extended beyond aesthetics; it also spurred dialogues about media, technology, and society, highlighting the growing integration of technological advancements within creative practices (Collopy, 2015).

Pioneering artists like Frieder Nake (1938–), Georg Nees (1926–2016), and Vera Molnar (1924–) were instrumental in advancing the integration of algorithmic and computational methods in art, fundamentally transforming artistic practices in the 1960s. Nake, a German computer scientist and artist, applied algorithms and computer programs to craft abstract visual compositions, notably seen in his *Grid Picture* (1965). This piece explores systematic structures, with its design rooted in a strict grid arrangement of intersecting horizontal and vertical lines, creating a precise and orderly pattern (Franco, 2022). The minimalistic color palette of the grid directs focus to its structural geometry, enhancing the sense of mathematical rhythm and compositional balance of the work. The repetitive grid elements reinforce a rhythmic, patterned effect, while the accuracy of the lines reflects Nake's commitment to computational logic and mathematical rigor in his artistic process.

Georg Nees also played a vital role in early digital art by using algorithms to generate complex visual forms and patterns. His work *Sine Curve Studies* (1969) employed mathematical sine functions to create intricate geometric shapes, with each line meticulously plotted to follow the flow of sine waves (Oliveira, 2021). This algorithmically generated composition features a series of sinuous curves that vary and evolve, resulting in a smooth, fluid appearance. The precision of the artist in plotting each curve contributes to a harmonious visual rhythm that characterizes his work. The systematic repetition and subtle variations of the sine curves give the artwork a dynamic, undulating quality. This meticulous and calculated approach underscores his dedication to computational aesthetics, demonstrating the possibilities of mathematical forms as an expressive medium in art.

Vera Molnar, a Hungarian-born artist, further explored the potential of algorithmic art by creating geometric abstractions through mathematical algorithms and systematic methods. In *Interruptions à recouvrements* (1969) (Fig. 3.5), her use of grids and intersecting lines generates a network of geometric forms that emphasizes complexity and depth (de Almeida, 2020). Her compositions are built

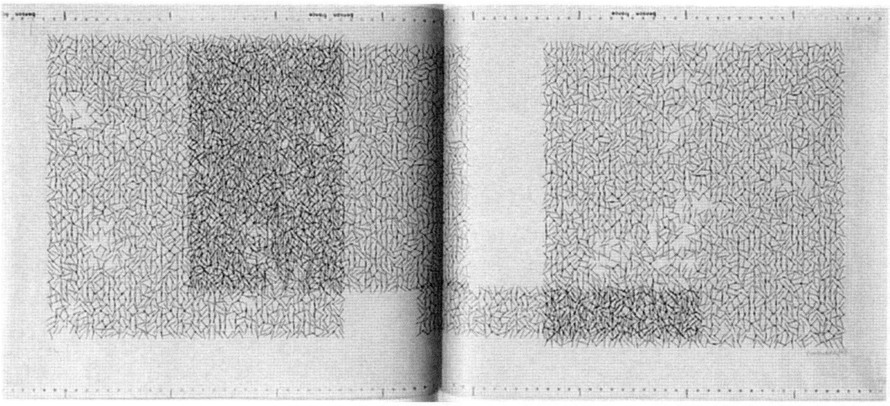

Figure 3.5 Vera Molnar, *Interruptions à recouvrements*, 1969 (Creative Commons License Zero).

on repetitive patterns governed by precise algorithmic rules, allowing for controlled intersections and overlapping lines that evoke visual intricacy. The structural basis of the work enables it to be displayed in various formats, from prints to digital screens, reflecting the innovative approach of the algorithmic artist to digital media. As one of the early adopters of computational techniques in art, her work bridges mathematical precision with aesthetic expression, encapsulating the exploratory spirit of the era.

Together, these artists set the stage for further innovations in digital art and computational creativity, demonstrating the potential of algorithms to expand traditional boundaries of artistic expression. Their work exemplifies an innovative approach to computer-generated art, redefining traditional methods of art-making and embracing the creative possibilities technology offers. The significance of these early pioneers lies in their success at bridging art and technology, showcasing the expressive potential of computer-generated imagery. Their creations not only questioned established artistic norms but also laid essential groundwork for the digital art movement. Through algorithmic and computational techniques, they opened avenues for future artists to explore technology as an integral part of artistic creation (Paul, 2023).

Around this time, a new form of art known as new media art emerged, encompassing a range of digital and multimedia practices. Key figures such as Nam June Paik and Jenny Holzer (1950–) significantly shaped this movement, pushing the boundaries of digital technology in their work. Paik was a pioneering force in integrating television sets, video projections, and electronic devices within his installations (Trail, 2021). His art challenged conventional ideas of what constitutes art and interrogated the evolving role of technology in society, inspiring further exploration within the new media art movement. The Koren-born artist's groundbreaking 1963 exhibition *Exposition of Music – Electronic Television* marked a pivotal moment in new media art, setting a precedent for the use of television and electronic imagery as an artistic medium (Paik, 1963). In this exhibition, Paik displayed television monitors that showcased manipulated, synthesized images, moving beyond the conventional use of television as a passive broadcasting tool. Instead, he transformed these monitors into active art forms, inviting viewers to engage with the screens as dynamic canvases. Paik's innovative use of synthesized images allowed for abstract patterns, visual distortions, and rapid shifts in imagery, which captivated audiences and challenged them to see technology in a new light. This approach not only redefined the role of television in art but also questioned the boundaries between mass media and personal expression.

The exhibition blurred the line between artist, machine, and audience, encouraging viewers to participate in the creative process. By co-opting a medium typically associated with mass communication and entertainment, he offered a commentary on the omnipresence of technology and its potential for both conformity and creativity. This exhibition demonstrated how electronic media could become a vehicle for artistic exploration rather than just information dissemination. His manipulation of television imagery invited audiences to question their perceptions of media and reality, laying the groundwork for future artists who would continue to experiment with electronic and digital forms in their art. His innovative use of technology in *Exposition of Music – Electronic Television* played a foundational role in establishing new media art as a distinct genre and underscored the expanding possibilities for art in the digital age (Hanhardt, 2006).

Another notable example of Nam June Paik's exploration of technology and spirituality is his 1974 installation *TV Buddha* (Fig. 3.6), which presents a Buddha statue facing a television monitor that displays a live feed of the statue itself. This seminal artwork juxtaposes ancient religious iconography with modern media, creating a reflective dialogue between Eastern philosophy and Western media culture (Holling, 2020). In the installation, a Buddha sits in a meditative pose, exuding an aura of serenity as it "watches" itself on the television screen. The live video feed produces a contemplative loop, where the Buddha gazes upon its own image, merging the real with the mediated (Lim, 2019). By positioning the television directly in front of the statue, Paik initiates a conversation on self-reflection and the influence of media in shaping self-perception.

The contrast between traditional and modern elements is intensified by the materials Paik employs. While the Buddha statue is often crafted from stone or other classical materials, the television monitor represents the cutting-edge technology of the 1970s, a time when electronic media was increasingly infiltrating daily life. This juxtaposition encourages viewers to reflect on the relationship between spirituality and mass media culture, as the warm glow of the television screen bathes the Buddha in light, creating a striking contrast between the illuminated figure and the surrounding darkness. This interplay between light and shadow enhances the meditative quality of the piece, adding layers of mystery and spirituality that invite deeper introspection.

The symbolism embedded in *TV Buddha* is profound, as Paik uses technology to reinterpret spirituality within a modern context. The Buddha's quiet contemplation of its own image emphasizes the influence of media on our understanding of self and tradition. The live video feed suggests a sense of perpetual observation, mirroring the constant surveillance present in contemporary media culture. Paik's installation not only challenges viewers to consider how media shapes personal and cultural values but also pushes the boundaries between art and media, creating an immersive,

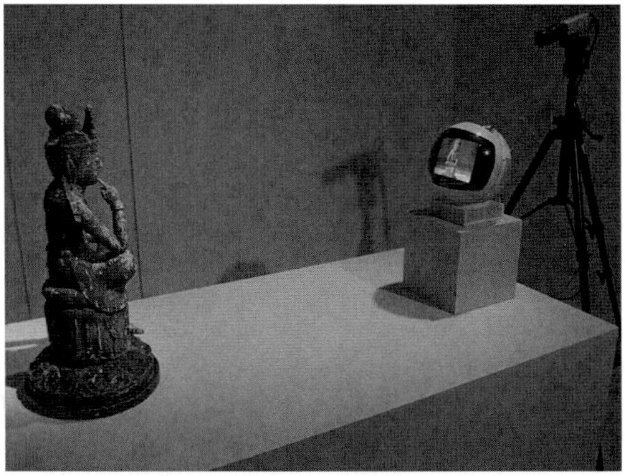

Figure 3.6 Nam June Paik, *TV Buddha*, 1974 (Creative Commons License 2.0).

thought-provoking experience that encourages self-reflection in a digitally saturated world.

The origins of video art in the 1960s and 1970s set the stage for a radical reimagining of art through technology, establishing foundational practices that continue to influence contemporary media arts. Early pioneers like Nam June Paik, Frieder Nake, Georg Nees, and Vera Molnar broke new ground by integrating television monitors, synthesized images, and algorithmic processes into their works, challenging traditional artistic boundaries and redefining the role of media in art. By incorporating video, computers, and interactive installations, these artists demonstrated the creative potential of technology to expand visual storytelling, self-reflection, and audience engagement. Their work bridged art with emerging fields like cybernetics and computer science, opening up new dialogues between human creativity and machine processes. The advances made during this era solidified video art as a legitimate art form and laid the intellectual and technological groundwork for future generations to explore digital, interactive, and AI-driven art, ultimately reshaping the landscape of modern and contemporary art.

3.3 Advances in Film and Videography Technology (1980s–1990s)

The 1980s and 1990s marked another transformative period in film and videography technology, characterized by rapid advancements that expanded creative possibilities and broadened access to production tools. This era witnessed the shift from analog to digital formats, a development that would forever change the landscape of filmmaking, video art, and digital media. Innovations such as digital video cameras made high-quality production more accessible, while the rise of CGI and special effects pushed the boundaries of what was visually possible. Software tools like Adobe Photoshop and Illustrator further revolutionized digital art, providing artists with powerful resources to manipulate and create images in unprecedented ways. Collectively, these technological breakthroughs empowered a new generation of filmmakers and artists to experiment with digital and visual effects, setting the stage for later advances in artificial intelligence and interactive media. This significance of the period lies in its foundational role in establishing digital art and CGI as central elements of modern visual storytelling and creative expression, paving the way for even more sophisticated AI-driven technologies in the following decades (LaRocco, 2018).

The early 1990s saw remarkable strides in merging artificial intelligence with visual media, marked by pioneering works that leveraged AI for creative exploration. In 1991 and 1992, Karl Sims won the Golden Nica award at Prix Ars Electronica for his 3D AI-animated videos *Panspermia* and *Liquid Selves,* which utilized artificial evolution to generate forms. Sims developed these animations by applying genetic algorithms, allowing random mutations of shapes to evolve and "select" forms based on visual appeal or novelty (Schöpf, 1991). This process not only introduced AI-driven aesthetics but also showcased the potential of evolutionary algorithms in creative fields. By 1999, Scott Draves further pushed the boundaries of AI in visual art with *Electric Sheep* (https://electricsheep.org/), a continuously evolving animation of fractal flames that adapted based on audience feedback (Draves, 1999). The work exemplified interactive digital art that learns from viewer input, creating a collaborative feedback loop that influenced the evolution of the visual experience.

These developments coincided with the establishment of large, publicly annotated image repositories, notably ImageNet, which would later be instrumental in training AI models for image recognition. These repositories provided the data sets necessary for algorithms to catalog and identify objects, laying the groundwork for the machine learning models that became prominent in the following decade. Together, these advancements demonstrated the early potential of AI-driven tools and interactive art, but more accessible tools were transforming videography at the same time.

One of the most impactful shifts in this era was the transition from analog to digital video technology, beginning with the introduction of digital video cameras like Sony's D1 in 1986. Unlike previous video cameras, which were large and cumbersome, digital cameras were more compact and affordable, allowing independent filmmakers greater flexibility and accessibility in producing high-quality films (Amyes, 2013). This technological shift sparked a surge in independent filmmaking, exemplified by Robert Rodriguez (1968–), who famously produced his 1992 film *El Mariachi* on a modest budget of $7,225. He shot the film on 16 mm film and later transferred it to digital format, showcasing how resourceful use of affordable technology could bring high-quality storytelling to the screen (Ricketts, 2014).

In fact, from the late 1980s and into the next decade, several video artists and independent filmmakers utilized digital video cameras to reshape storytelling and artistic expression, each bringing unique insights and innovations that pushed the boundaries of the medium. The Swiss-born Pipilotti Rist (1962–) used digital video in *I'm Not The Girl Who Misses Much* (1986) (https://youtu.be/hjvWXiUp1hI?si=ZpblOZ466AYeGE4J) (Fig. 3.7) to explore themes of female sexuality through a playful, almost chaotic remix of pop culture, fantasy, and everyday life. The merging of pop music with distorted visuals and electronic

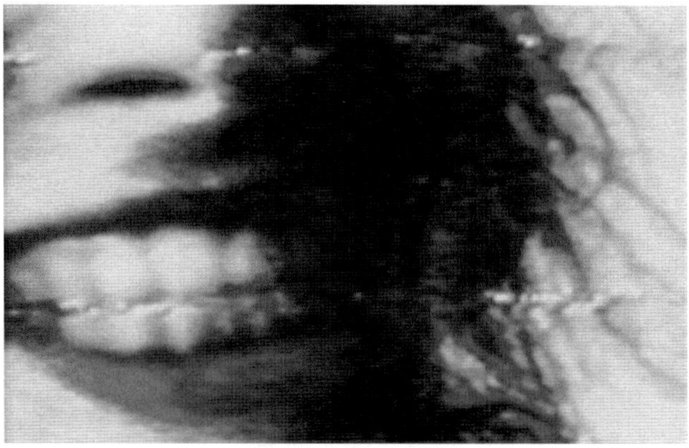

Figure 3.7 Pipilotti Rist, *I'm Not the Girl Who Misses Much*, 1986 (Creative Commons License 4.0).

manipulation allowed Rist not only to contribute to feminist discourse in video art but also highlighted how digital technology could amplify personal and political expressions. Her work underscored the potential of video as a medium for addressing complex social themes in a way that is both engaging and provocatively immersive, expanding the impact of video art on contemporary culture (Watlington, 2022).

In the 1990s, *Gummo* (1997) directed by Harmony Korine (1973–) demonstrated the versatility of digital video in independent filmmaking with its raw, improvisational style and open-ended narrative structure. Shot with digital cameras, *Gummo* offered a fragmented view of small-town American life, allowing audiences to impose their interpretations and reflect on its experimental approach. This flexibility and freedom from conventional narrative allowed Korine to create a film that felt unfiltered and authentic, marking a significant departure from traditional cinema and influencing a wave of independent filmmakers to embrace digital technology as a way to experiment with form and content (Sklar, 2001).

Similarly, *The Cremaster Cycle* (1994–2002) (https://www.youtube.com/watch?v=6xWtS9HsP4U?) by Matthew Barney (1967–) broke ground in both scope and thematic complexity, blending film, sculpture, and performance art to explore myth, identity, and the human body. His use of digital technology enabled the creation of visually elaborate scenes and layered storytelling, making it one of the most ambitious works in avant-garde cinema. The *Cremaster Cycle* became emblematic of how digital tools could expand not only the visual possibilities but also the thematic depth of experimental film, encouraging other artists to explore hybrid approaches that fused different forms and mediums (Frichot, 2015).

Each of these works contributed to a broader evolution within video art and independent filmmaking, highlighting the effects of digital video on visual storytelling, accessibility, and artistic exploration. From Rist's technological manipulations to Barney's mythological explorations, these artists demonstrated how digital tools could transcend traditional art forms, making space for a new generation of creators to push the limits of both narrative and visual experimentation. Their pioneering works underscored the potential of digital video to deepen the emotional and intellectual engagement of audiences, cementing the significance of this period in the history of film and video art.

Alongside advancements in digital cameras, the development of CGI revolutionized the visual effects industry. Films like *Tron* (1982) and *Jurassic Park* (1993) were not just groundbreaking in their own right; they represented the first successful integration of computer-generated imagery into mainstream cinema. In *Jurassic Park*, Industrial Light & Magic (ILM) set a new standard by combining CGI with animatronics, seamlessly blending full-scale puppetry with digitally animated dinosaurs to achieve lifelike movement and texture. Visual effects supervisor Dennis Muren and animator Phil Tippett pushed animation techniques beyond the limitations of go-motion and stop-motion, crafting a level of realism previously unattainable. By using Softimage 3D software to render digital elements and integrating them with physical models from Stan Winston Studio, the team created a T-Rex sequence that captivated audiences, laying the groundwork for CGI-driven creatures in live-action films (Shay and Duncan, 1993). This blend of practical and digital effects set the industry on a path toward increasingly sophisticated digital characters, inspiring a shift toward CGI that would soon become central to blockbuster filmmaking.

In the latter half of the 1990s, *The Matrix* (1999) introduced audiences to another leap in visual effects with its "bullet time" innovation, further advancing the capabilities of digital compositing and visual storytelling. Visual effects supervisor John Gaeta employed a precise array of 120 still cameras alongside motion picture cameras, arranged in a circular formation, to capture protagonist Neo dodging bullets in a surreal, time-warped sequence. Custom interpolation software developed by Manex Visual Effects generated intermediate frames, creating smooth motion through space while distorting time—a technique that would become iconic and heavily parodied. The bullet time effect, achieved through digital compositing with Flame and Inferno software, enabled a level of spatial and temporal manipulation that forever changed action cinematography and established new paradigms for integrating CGI into narrative-driven sequences (Rehak, 2007). This innovation highlighted the increasing reliance on custom software development for specific cinematic needs, showcasing digital technology's potential to redefine visual storytelling.

Motion capture (MoCap), gaining traction in the late 1990s and early 2000s, bridged live-action and animation, capturing actor movements with unprecedented precision and enabling digital characters to mimic real-life gestures and expressions. This technique played a pivotal role in *The Lord of the Rings* trilogy (2001–2003), where actors' performances were seamlessly blended with CGI characters, adding a depth of realism to Gollum and other digital figures. As MoCap advanced, filmmakers faced new compositing and editing challenges, requiring animators to integrate captured data into CGI frameworks, merging digital and human performances in ways that expanded the boundaries of visual realism (Bregler, 2007). The technological evolution continued with James Cameron's *Avatar* (2009), which pushed MoCap to new heights through innovations in facial capture and virtual camera systems. Led by effects supervisor Joe Letteri, Weta Digital developed head-mounted cameras to capture even the most minute facial expressions, generating massive data sets—up to 1.4 gigabytes per frame. Volumetric lighting systems, advanced muscle simulation software, and a 360-degree volumetric capture stage established new standards in character animation and environmental interaction, creating photorealistic digital characters that were lifelike yet fantastical (Katz and Ellis, 2024).

These technological advancements in CGI and MoCap created the conditions for future AI-driven effects, as filmmakers increasingly relied on data-heavy, programmable digital elements that could adapt to the demands of complex scenes. The developments in software, such as volumetric capture and real-time rendering through game engines like Unreal, enabled directors to view completed digital effects during principal photography, further merging physical and digital workflows (An, 2022). This era's innovations in CGI, special effects, and digital compositing marked a turning point, allowing filmmakers to realize creative visions that had previously been unimaginable. The legacy of these breakthroughs is evident in contemporary blockbusters, where AI and real-time engines drive everything from digital environments to virtual actors, underscoring how this period laid the foundation for the sophisticated digital effects and AI-driven artistry shaping today's cinematic landscape.

As CGI and MoCap technologies transformed the visual effects landscape in film, the development of software tools like Adobe Photoshop and Illustrator during the same period began reshaping the field of digital art and visual media. Introduced in the late 1980s, these software programs offered artists new capabilities to create, manipulate, and enhance images in ways that had previously been limited to physical media. Adobe Photoshop (Fig. 2.2), released in 1988, provided tools for image manipulation that allowed artists to blend multiple images, adjust colors and textures, and experiment with digital effects, expanding the creative possibilities for photographers, illustrators, and designers alike (Sugiarto et al., 2021). Similarly, Adobe Illustrator, launched in 1987, revolutionized the creation of vector graphics, enabling scalable, precise illustrations that retained their quality at any size. With its digital brushes, filters, and design tools, Illustrator allowed artists to develop intricate digital art, blurring the line between traditional illustration and digital creation (Wang, 2021).

These software innovations marked a turning point, empowering artists to push the boundaries of artistic expression through experimentation and iteration. Photoshop's flexibility allowed artists to explore new forms of visual storytelling by blending elements from photography, painting, and digital composition. Illustrator's vector-based approach facilitated the creation of clean, scalable art that could be applied across various mediums and formats, from print to digital displays. Together, these tools enabled a seamless fusion of digital and traditional techniques, allowing artists to challenge conventional notions of art and unlock unprecedented creative freedom. This period saw the emergence of a digital art movement, where the accessibility and versatility of tools like Photoshop and Illustrator began to popularize art production, making it possible for a wider range of creators to explore and express their visions without the constraints of traditional art materials (Zhang, 2020).

As digital tools redefined visual media, parallel developments in artificial intelligence research, particularly in language modeling, would later expand AI's role in creative fields. The concept of N-gram language modeling, prominent in the 1960s and 1970s, laid the foundation for early text generation by enabling AI systems to predict word sequences based on learned patterns. This breakthrough in understanding sequential data gained momentum in the late 1990s with the introduction of Long Short-Term Memory (LSTM) networks, which addressed the challenge of capturing long-term dependencies in data. By introducing memory cells and gating mechanisms, LSTMs made it possible for AI to learn complex patterns over extended periods, enhancing applications in speech recognition, machine translation, and text generation (MA et al., 2015). These advancements in sequence modeling paralleled the evolving landscape of digital art, setting the stage for future role of AI in generating not only text but also visual media and interactive experiences.

As AI-based tools began integrating into visual media creation, these foundational technologies from the 1980s and 1990s paved the way for a digital revolution in art and media, bringing us closer to the sophisticated, AI-driven creativity that defines today's digital landscape. The groundwork laid by software like Photoshop, which enabled seamless manipulation of images, and Illustrator, which made vector-based design scalable and precise, helped bridge traditional and digital media. These advancements also offered a conceptual foundation for future AI-driven tools, demonstrating the potential of technology to not only support but actively transform artistic expression. The technological and conceptual progress of this era created a fertile environment for innovation, inspiring exploration in AI's potential within creative fields. With developments like LSTMs allowing for more complex sequence generation, AI applications expanded into natural language processing, dialogue systems, and more. Together, these innovations in digital art and AI modeling paved the way for increasingly sophisticated generative models and have become instrumental in shaping the contemporary landscape of AI-driven artistry and media production.

3.4 The Rise of AI in Video Art (2000–2010)

The first decade of the new millennium marked a crucial turning point for video art, as advancements in digital technology and artificial intelligence began to reshape the creative landscape. Building on the foundations laid by digital tools like Adobe Photoshop and Illustrator in the 1990s, artists increasingly incorporated AI and machine learning into their work, allowing for unprecedented experimentation and interactivity. This period saw the rise of generative and algorithmic art, where artists used AI to create dynamic visual experiences that could adapt and respond to viewers or environmental inputs. Video art installations became more immersive and complex, reflecting the growing sophistication of software and the integration of machine learning in creative processes. By the early 2000s, AI-driven techniques like facial recognition and computer vision enabled artists to create video works that reacted to audience movements and expressions, blending virtual and physical spaces in ways previously unimaginable. These technological strides not only expanded the tools available to artists but also challenged traditional notions of authorship, control, and artistic agency, setting the stage for AI as a transformative force in video art throughout the decade.

The 2000s were notable for the widespread acceptance and acknowledgment of new media, particularly video art, within major art institutions. As digital art gained prominence, museums and galleries began to exhibit and acquire digital and video artworks alongside traditional forms. Institutions like the Museum of Modern Art (MoMA) and the Whitney Museum of American Art spearheaded this shift, offering dedicated exhibitions and acquisitions that established the legitimacy of digital art within the fine arts realm (Morigi, 2004). Biennials and festivals, including the Whitney Biennial and Electronic Arts Intermix, further showcased video art and new media, providing critical platforms for digital artists and encouraging the integration of these forms into mainstream art discourse. The 2000s saw digital art recognized not merely as a technical novelty but as a vital part of contemporary art, bridging technological innovation and artistic expression.

Jenny Holzer (1950–) was a prominent figure in this era, continuing her work into the new millennium with provocative installations that used digital technology to convey powerful social and political messages. Holzer's LED text installation *7 World Trade Center* (2006) (Fig. 3.8) exemplifies her ability to engage with public spaces through site-specific art that resonates with its surroundings. This piece, displayed on the facade of the 7 World Trade Center building in Lower Manhattan, featured scrolling LED text that commanded attention and interacted directly with the architectural space (Holzer, 2021). The installation utilized large, bold letters projected onto the surface of the building, creating a dynamic, ever-changing visual experience that invited viewers to reflect on the significance of the site. The LED technology contributed a modern and luminous element, with bright, vibrant colors contrasting sharply against the surrounding architecture, amplifying the visibility and impact of the message.

The text in *7 World Trade Center* played a vital role in its meaning, with Holzer often incorporating phrases that addressed complex social, political, and cultural themes. Situated at the World Trade Center site, the scrolling text could evoke themes of memory, resilience, and renewal, touching on the tragedy and rebuilding efforts following the events of September 11, 2001. Holzer's choice of LED technology added an urgency to the piece, as the moving text compelled viewers to engage with the message in real-time, underscoring the work's dynamic relationship with its surroundings (Luger, 2015). Through the use of scrolling text, vibrant LED lights, and a thoughtful site-specific approach, Holzer created an artwork that not only highlighted social and cultural themes but also honored the resilience of a place marked by loss and recovery. The piece invited introspection and dialogue, reinforcing the power of digital technology to create visually striking, emotionally resonant public art.

Along with conceptual artists using digital media for their work, advances in video art continued to build on previous innovations. During the first decade of the new millennium, video artists adopted a range of approaches to explore the intersections of digital media, time, emotion, and identity. Among these, *The Clock* (2010) (https://youtu.be/NZ5I0H-WoKA) by Christian Marclay (1955–) stands out as an iconic example. This 24-hour video montage compiles thousands of film and television clips featuring clocks and watches, edited so the clips reflect the real

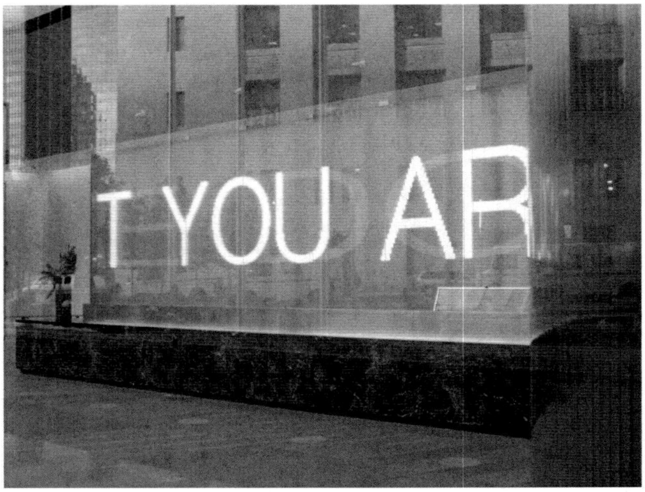

Figure 3.8 Jenny Holzer, *7 World Trade Center*, 2006 (Creative Commons License Zero).

time of the day, effectively functioning as a timepiece. The painstaking research and editing on the part of the artist created a work that traverses various narratives, settings, and moods within minutes, highlighting the influence of time in cinema and daily life. Through the weaving together of diverse cinematic moments around a shared theme, Marclay provides viewers with a unique, immersive experience that challenges the perception of time and continuity in both film and reality (Levinson, 2015).

Other creatives used this new technology to elicit themes from the history of art. One such example comes from Bill Viola (1951–2024) and his *The Quintet of the Astonished* (2000) (https://youtu.be/nSlMgPXOkmk), which offers a deeply emotional and painterly experience, using extreme slow motion to capture the shifting expressions of five actors as they experience a range of emotions. Viola, a pioneer in new media art, creates an almost devotional experience by emphasizing each minute detail of the faces of the actors, reflecting his influences from medieval and renaissance art. This focus on human expression and suffering speaks to universal experiences of grief, catharsis, and resilience. His work often references traditional art forms, yet it incorporates cutting-edge video techniques that evoke a meditative state, inviting viewers to contemplate the depth of human emotion and the effect of time on perception (Viola, 2013).

While artists like Bill Viola used new technology to reinterpret the themes and compositions of Renaissance altarpieces, bringing a timeless subject matter into contemporary form, others embraced the evolving gallery spaces, utilizing multi-screen installations to create immersive experiences that directly engaged the audience and encouraged active participation. In *Play Dead: Real Time* (2003) (https://youtu.be/Q-XD6fuf0ho), Douglas Gordon (1966–) explores themes of control, vulnerability, and spectacle through an unusual subject—a circus elephant filmed in a stark white gallery space, performing commands to lie down and rise. Filming in slow motion and presenting the footage from multiple perspectives,

Gordon places the viewer in a reflective position, prompting questions about power and freedom. The movement of the elephant in the gallery space evoke empathy, as it follows commands that emphasize its physicality and seeming subservience. This multi-perspective approach underscores the critique of spectacle and the human inclination to observe and control nature, themes that resonate within contemporary society and media (Gordon, 2003).

Artists like Douglas Gordon explored multiple screens to actively engage and immerse audiences within layered narratives, while others, such as Steve McQueen (1969–), used the medium in installations to evoke themes of emptiness and absence, capturing the subtle tensions and silences within physical spaces. His *Giardini* (2009) (https://youtu.be/UlG1wnCpdxQ) illustrates this, taking a contemplative approach, capturing the deserted pavilions of the Venice Biennale during the off-season in a two-channel video installation. Filmed in a quiet, wintry setting, the piece juxtaposes the grandeur of the art world with moments of emptiness and solitude. By including glimpses of everyday life—stray dogs, rain, and subtle sounds—McQueen emphasizes the contrast between art's vibrant presence during the Biennale and its haunting absence in its off-season. This work reflects on the cycles of visibility and neglect within the art world, invoking themes of impermanence, memory, and cultural significance (Robinson, 2015).

The quiet introspection found in the work of McQueen transitions into the overwhelming cacophony of life amidst modern technologies in the creations of artists like Ryan Trecartin (1981–), whose chaotic, multi-layered narratives reflect the frenetic energy and fractured identities of the digital age. Works like *I-BE AREA* (2007) (https://youtu.be/V27rH6b5ub4) embody the chaotic, layered nature of internet culture with its blend of live-action and digital effects. Characterized by a fragmented narrative, frenetic pacing, and jargon-laden dialogue, the work captures the aesthetics of digital media and the rapidly evolving nature of online identities. Through eccentric characters and DIY-style sets, Trecartin presents a vision of contemporary identity that is fluid, interconnected, and constantly shifting. This work explores the impact of digital media on personal expression and the formation of identity, reflecting the internet's pervasive influence on culture and communication (Åkervall, 2015).

Finally, Shirin Neshat (1957–) exemplifies a group of video artists who have harnessed the medium to deeply consider human stories, using film and video as powerful tools to explore themes of identity, political struggle, and cultural memory. The film *Women Without Men* (2009) (https://youtu.be/UlG1wnCpdxQ) addresses such complex issues of gender and political struggle within a powerful narrative structure. Adapted from Shahrnush Parsipur's novel, the film examines the lives of Iranian women during the 1953 coup in Tehran, using striking chiaroscuro lighting and controlled color to heighten the sense of tension and isolation. The work of Neshat combines elements of magical realism and historical narrative to evoke the personal and political turmoil experienced by her characters, illuminating cultural dynamics often underrepresented in mainstream media. Her visually arresting style reflects the merging of traditional Iranian art with contemporary filmmaking techniques, bringing social and political issues into sharp focus (Neshat et al., 2009).

The work of these video artists exemplifies the creative possibilities unlocked by digital and multimedia tools in the early 2000s. Through techniques such as slow-motion, real-time synchronization, multi-perspective shots, and interactive elements, each artist expanded the boundaries of video art to explore themes of time, identity, emotion, and societal dynamics. These advancements in digital media not only enriched artistic expression but also marked the decade as a significant period of growth and experimentation in video art. While these video artists investigated how new technologies could provide deeper insights into cultural experiences and the essence of human existence, Hollywood and global cinema were concurrently integrating various new technologies to enhance storytelling and elevate special effects in mainstream cinema.

In fact, the decade brought substantial advancements in digital filmmaking technologies, with directors exploring new methods to push the boundaries of cinematic storytelling and visual style. One such approach that foreshadowed big-budget spectacles can be seen in the work of director Robert Rodriguez. In *Sin City* (2005), Rodriguez utilized the creative potential of digital backlot techniques by filming actors entirely against green screens, enabling the environments to be constructed entirely in post-production. The approach enabled the *El Mariachi* director to create a stylized, high-contrast visual aesthetic that closely resembled the graphic novel origins of the story, blending live-action with a comic book feel. The digital backlot provided freedom to manipulate every visual element, from lighting and perspective to color schemes, in ways that traditional sets could not achieve. This technique allowed for greater control over the atmosphere and artistic direction, creating a distinctive, surreal environment that brought the graphic novel by Frank Miller (1957–) to life. Work on the film highlighted how digital post-production could give filmmakers new creative tools to construct unique visual worlds, setting a precedent for future stylized films (Pallant, 2007).

Meanwhile, *Star Wars: Episode II – Attack of the Clones* (2002) marked a historic moment in cinematography by being one of the first major films shot entirely on digital cameras. While George Lucas (1944–) had established himself as a director that pushed the boundaries of special effects since *Star Wars: Episode IV-A New Hope* (1977), this shift from analog film to digital cinematography was significant not only because it showcased the technological capabilities of digital cameras but also because it signaled a major shift in the industry approach to film production. Digital cinematography offered a more efficient and flexible alternative to traditional film, providing faster workflows and the ability to review shots immediately on set. For Lucas, the use of digital cameras facilitated the extensive CGI work in *Attack of the Clones*, as digital footage could be seamlessly integrated with computer-generated environments and characters. This landmark decision influenced countless filmmakers to adopt digital technology, gradually establishing digital cinematography as a standard practice in the industry and opening doors for complex visual effects that relied on digital elements (Lipton and Lipton, 2021).

Along with digital cinematography, other special effects advances would become industry standard. An excellent example of this comes by way of *The Curious Case of Benjamin Button* (2008) where *Fight Club* director David Fincher (1962–)

showcased groundbreaking digital aging and de-aging techniques that allowed for a seamless transformation of the appearance of actor Brad Pitt throughout the film. Fincher's team used advanced CGI and motion capture technology to create realistic age transitions, a feat that was previously challenging to achieve convincingly. The combination of 3D facial scans, CGI, and performance capture, allowed Fincher to depict Benjamin Button aging backward, blending actor performance with digital alterations that maintained the essence of the actor across different stages of life. This innovation set a new standard for character transformations, providing filmmakers with a powerful tool to explore complex narratives involving age, identity, and time. The success of this technique paved the way for further exploration of digital aging effects, which have since become more refined and widely used in film production (Loock, 2021).

The Adventures of Tintin (2011) took digital filmmaking further by employing performance capture technology to animate characters, blending live-action performances with digital animation. Director Steven Spielberg (1946–) oversaw the performance capture, which allowed actor movements, facial expressions, and body language to be digitally recorded and applied to animated characters, creating a fusion of realistic human gestures within a digitally constructed world. This technique enabled Spielberg to bring the classic comic book characters to life with a level of nuance and realism that traditional animation could not achieve. The use of performance capture in *Tintin* emphasized the growing trend of merging animation with live-action elements, creating a new space for hybrid storytelling. The success with this technology here contributed to the increasing popularity of performance capture in blockbuster films, especially for characters requiring high levels of expressiveness within fantasy or animated contexts (Carter, 2019).

Additionally, Michael Mann's (1943–) *Collateral* (2004) utilized digital cinematography to capture night scenes with unprecedented clarity and realism, showcasing the unique capabilities of digital cameras in low-light conditions. Traditional film cameras often struggled to capture crisp details in dimly lit environments, but the use of digital cameras in this instance allowed Mann to reveal the subtle textures and atmosphere of Los Angeles at night. This approach brought an authenticity to the setting of the film, enhancing the suspense and immediacy of the narrative. The clarity of digital footage in low light emphasized the contrast between light and shadow, creating a vivid sense of depth and realism that would influence future filmmakers working with urban landscapes. Work on *Collateral* demonstrated that digital cinematography could achieve a new level of visual realism, especially in challenging lighting conditions, further solidifying digital technology's role in the industry (Rayner, 2013).

Together, the technological innovations of the 2000s by both filmmakers and video artists demonstrated the transformative potential of digital tools in advancing storytelling and visual aesthetics. Directors like Robert Rodriguez, George Lucas, David Fincher, Steven Spielberg, and Michael Mann pioneered digital backlots, CGI integration, performance capture, and digital cinematography, reshaping the possibilities within cinema Meanwhile, video artists such as Christian Marclay, Bill Viola, and Shirin Neshat were exploring digital media in multi-screen installations, real-time synchronization, and interactive elements, pushing video art to new expressive heights. These artists employed digital tools to investigate complex themes of time, identity, and cultural memory, often inviting viewers into immersive, introspective experiences that paralleled the cinematic ambitions of the film industry.

As both filmmakers and video artists experimented with digital techniques, they expanded the scope of creative expression in unique ways. In the film industry, technological advancements enabled fresh approaches to storytelling, from creating entirely virtual worlds to blending live-action with digital environments. In parallel, video artists were innovating within gallery spaces, using new media to capture the human experience, often creating installations that invited direct audience engagement. Together, these advancements across film and video art set the stage for even more sophisticated digital and AI-driven effects, forever altering the landscape of visual storytelling and artistic expression in the years to come.

3.5 Interactive Video and GAI Integration (2010s–Present)

The influence of AI on the visual arts became especially significant in the 2010s, marking the start of a new era in image generation and manipulation. A pivotal breakthrough came in 2014 with the development of Generative Adversarial Networks (GANs), which transformed the field by enabling the creation of realistic, high-quality images. Developed by Ian Goodfellow and colleagues at Université de Montréal, GANs introduced a type of deep neural network capable of learning to replicate the statistical distribution of input data, including images. The GAN framework consists of two components: a "generator," which creates new images, and a "discriminator," which evaluates these images to determine their success. This dynamic interplay allows the network to continuously refine its outputs (Goodfellow et al., 2020). Unlike previous algorithmic art methods, GANs can learn to produce any given style by analyzing large datasets, making them a major evolution in the field (Gauthier, 2014). This technology has since become an invaluable tool for artists and designers, who can now generate visually compelling and diverse images that extend the possibilities of traditional artistic creation.

Another important innovation in 2014 was the introduction of Variational Autoencoders (VAEs) (Figueira and Vaz, 2022). VAEs work by compressing data into a latent space—a simplified representation that captures essential features—and then decompressing it to reconstruct or generate new images. The capability to learn latent data representations facilitates high-quality image creation, as VAEs can capture the core structure and style of data while retaining flexibility. Through modeling complex distributions within data, VAEs enabled artists to manipulate these latent spaces, producing variations of images or entirely new compositions by adjusting specific elements or blending styles. The process opens up a vast range of artistic possibilities, as artists can explore subtle shifts in color, form, or texture with unprecedented control. Furthermore, VAEs make it possible to create images that blend multiple influences or styles, allowing for a fusion of elements that might not traditionally coexist in a single composition. The adaptability of VAEs has thus made them invaluable for artists seeking to push the boundaries of digital creativity, offering depth, variety, and a degree of creative freedom that traditional methods cannot achieve (Wei et al., 2020).

A major factor driving interest in GAN models was Google's DeepDream, introduced in 2015. DeepDream used a convolutional neural network to identify and amplify patterns within images through a process called algorithmic pareidolia, resulting in visuals with a surreal, dream-like appearance (Fig. 3.9) (Bhardwaj et al., 2023). Shortly after, in 2016, StackGAN made significant advancements in AI-driven image synthesis by employing two separate GANs to generate high-resolution images from text descriptions, setting a new standard for detail and realism in AI-generated visuals. The StackGAN framework operates in two stages: first, it generates a rough, low-resolution version of an image, which is then refined into a high-resolution version by a second GAN, adding intricate details and improving overall quality. This two-step process allowed for the precise rendering of complex features based on simple text inputs, giving artists the ability to transform abstract concepts or detailed descriptions into vivid, accurate images (Zhang et al., 2018). The control and fidelity provided by StackGAN empowered artists and designers to use descriptive language to realize complex visual ideas, bridging the gap between conceptual visualization and high-quality, concrete outputs.

The ability of AI to generate images in specific styles opened up the unprecedented possibility of reproducing the unique visual language of an artist without their direct involvement. This development meant that AI could theoretically emulate the approach of any artist by analyzing a sufficient dataset of their works. A high-profile example of this occurred in 2016 with the project *The Next Rembrandt*, undertaken by the advertising agency J. Walter Thompson Amsterdam. The team trained a model on a dataset comprising of 346 works attributed to Rembrandt (1606–1669), examining details such as brushstroke patterns, use of light, and composition style. Using these insights, the model was able to synthesize a new painting (Fig. 3.10) that closely resembled what Rembrandt himself might have created had he continued working (Sovhyra, 2021).

The project demonstrated not only the technical potential of AI in recreating historical styles but also raised intriguing questions about originality, authorship, and the nature of artistic legacy. By generating an entirely new "Rembrandt" painting, the project illustrated how AI could extend the legacy of an artist beyond their lifetime, preserving and continuing a specific artistic vision. At the same time, it sparked discussions around the ethical implications of using AI in art: Does an AI-created painting have the same value or authenticity as one created by the original artist? And to what extent should AI-generated works be considered original creations? Such projects led Mezei (2021) to call for more AI-pessimism, though they do highlight both the technical capabilities and the philosophical complexities of AI in the art world, as it continues to push the boundaries of how we define art, creativity, and artistic ownership.

Building on this, 2017 saw the emergence of StyleNet, a model that introduced the ability to generate captions for images and videos in different stylistic tones. StyleNet was notable for exploring the interplay between text and image, enabling AI to create captions that aligned with the aesthetic or emotional tone of the visual

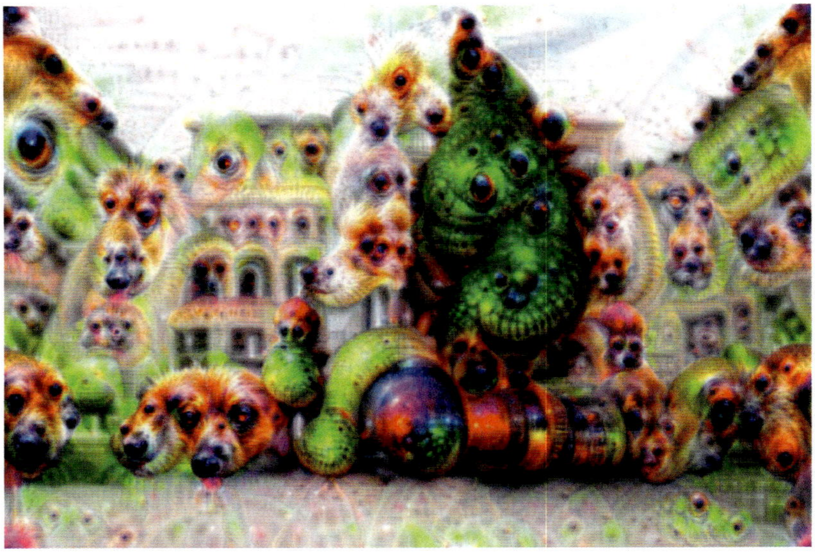

Figure 3.9 Leo Panthera, *DeepDream Schillergym*, Deep Dream, Google, 2021 (Creative Commons License Zero).

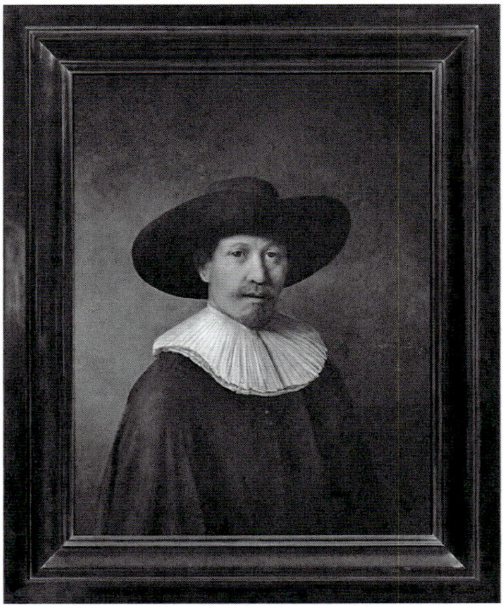

Figure 3.10 J. Walter Thompson Amsterdam, *The Next Rembrandt*, 2021 (Creative Commons License Zero).

content (Gan et al., 2017). For artists, the model offered a new layer of creative engagement, as they could experiment with text that either complemented or contrasted with the visual themes of their work, enhancing narrative depth. Through the synthesis of language styles that matched specific artistic intentions, this early generative model gave creators the tools to explore visual and textual storytelling simultaneously. This model thus enriched the creative process, allowing artists to explore how language could influence the interpretation of an image and vice versa, further integrating the role of AI in multi-dimensional art.

Another significant example of AI-generated art is *Edmond de Belamy* (Fig. 3.11), a portrait created by the Parisian arts collective Obvious in 2018. The team trained a GAN using an open-source algorithm developed by Dublin-born artist Robbie Barrat (1999–), feeding it a dataset of 15,000 portraits from the fourteenth to the nineteenth centuries sourced from the online art encyclopedia WikiArt. The result was a haunting, slightly distorted portrait that blended traditional portraiture techniques with the distinctive distortions and visual interpretation of the AI aesthetic (Goenaga, 2020). The work became a sensation when a print of it sold at Christie's for $432,500, far exceeding its initial estimate and marking one of the first times AI-generated art entered a major auction house. This sale ignited conversations around the value, authorship, and authenticity of AI-generated art, bringing it to the attention of the mainstream art market and elevating its status within the art world (Stephensen, 2019).

That same year, Artbreeder launched as the first widely accessible online AI image generator, making it possible for users to generate and customize images with ease. Artbreeder, based on StyleGAN and BigGAN models, allowed users to blend images, adjust parameters, and create unique visuals by combining multiple influences. Its user-friendly interface and extensive customization options helped widen the scope of AI art, allowing users without technical expertise to experiment with AI and create visually compelling pieces. This platform marked a shift in the accessibility of AI tools, as creative experimentation with AI was no longer limited to artists with advanced technical skills but could be explored by a broader community (Bertelsen and Short, 2023).

AI art soon began to gain recognition and accolades. In 2019, artist Stephanie Dinkins received the Creative Capital Award for *Not the Only One* (NTOO), an ambitious project described as a multigenerational memoir of a Black American family, narrated by an AI of evolving intellect. Dinkins trained the AI to learn stories from her family members, using it as a means to explore cultural memory, identity, and legacy within an interactive digital framework. *NTOO* represented a major step forward in storytelling, highlighting how AI could be used to preserve and retell narratives from underrepresented communities, thus blending personal history with cutting-edge technology. Such recognition encouraged a new generation of creatives to adopt the technology in their own work.

One such artist is Memo Akten (1975–), a Turkish artist based in London, who has delved deeply into the artistic applications of deep learning models. His 2021 work *Distributed Consciousness* (https://www.memo.tv/works/distributed-consciousness/) explores the cognition of the octopus, using AI to simulate synthetic alien intelligences. This project pushes AI-generated art into uncharted territory, raising profound questions about intelligence and the intricate connections between human and machine cognition. Through deep learning, Akten seeks to engage audiences in contemplating the complexity and diversity of consciousness beyond human experience, emphasizing the broad and often mysterious nature of intelligence (Berio

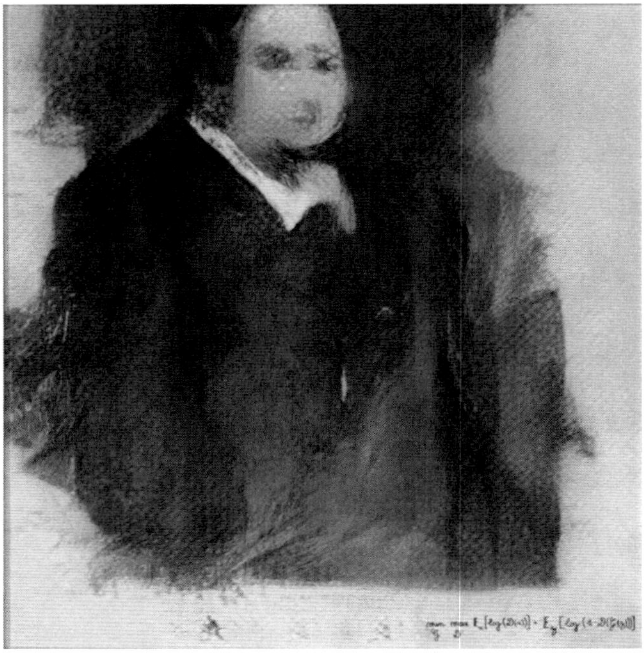

Figure 3.11 Obvious, *Portrait of Edmond de Belamy*, 2018 (Creative Commons License Zero).

et al., 2017; Deterding et al., 2017). His work exemplifies how AI can inspire artists to investigate themes that extend beyond human-centric perspectives, thus expanding the philosophical scope of contemporary art.

The rise of robotics in the art world has opened new avenues for creativity, with artists using this technology to explore collaborative and interactive forms of expression. Sougwen Chung, a multidisciplinary artist born in China, raised in Canada, and currently based in London, has made significant contributions to this field. As winner of the Lumen Prize, her work centers around a series of robots she developed called *D.O.U.G. (Drawing Operations Unit: Generation_1)* (2015) (https://sougwen.com/project/drawing-operations), which use AI and recurrent neural networks to learn and mimic her drawing style. Chung's integration of AI-driven robotics enables a dynamic, interactive creative process, as her robots respond to and collaborate with her in real-time. Through live performances and installations, Chung engages in a unique dialogue between human and machine, challenging traditional ideas of authorship and redefining creativity (Schnugg, 2019). By working alongside these robots, Chung explores the concept of embodied AI and the potential for collaborative partnerships between humans and machines. Her work raises critical questions about technology's role in the arts, the boundaries of creative expression, and the evolving nature of artistic collaboration in a technologically driven world.

While some artists fully embrace advanced technology, others blend traditional techniques with AI, creating a hybrid approach that bridges analog and digital art forms. One such artist is Linda Dounia, a Senegalese artist and curator born in

1994, whose work combines GANs with traditional materials. Dounia trains AI models on her own art, creating pieces that reflect both human spontaneity and the structured processes of AI. Her 2022 NFT project *Dust is Hard to Breathe* (https://nft.artsy.net/artwork/linda-dounia/dust-is-hard-to-breathe/6/) showcases curated outputs from a GAN, engaging with themes of identity, resistance, and the limitations of AI-generated art (Boucher, 2023). Her exploration of identity is central to her work, as she examines the intersection of human expression and the constraints inherent in AI. Through the use of GANs, she produces artworks that challenge conventional definitions of artistic authorship and push the boundaries of what creativity means in the context of AI. Through this synthesis of digital and physical mediums, Dounia provides a thought-provoking commentary on the evolving relationship between tradition and technology in contemporary art.

Some artists are using AI to interrogate the biases embedded within algorithmic systems, raising important questions about representation and inclusivity in technology. British media artist Jake Elwes (1993–) employs AI to explore how computer systems are trained and the biases that can arise from these processes. In his video piece *Zizi – Queering the Dataset* (2019) (Fig. 3.12), Elwes integrates the faces of drag performers into existing facial recognition datasets to confront the narrow representation often found in these algorithms. Through the deliberate modification of datasets, the artist exposes the inherent limitations and biases present in AI, drawing attention to the real-world implications of such biases. His work highlights the crucial need for diverse and inclusive datasets in AI development, advocating for fair and equitable representation in technological advancement (Voto, 2022). Such an exploration of algorithmic bias is not only a critique but also an empowering gesture for marginalized communities. Through his art, he aims to amplify voices and identities that have been historically underrepresented or misrepresented within these systems. The work underscores the idea that technology, when consciously designed, can be a powerful tool for social change, promoting inclusivity and equity within AI and its applications.

These examples demonstrate the viability of generative visual tools to assist artists as a new medium, but they remained the domain of specialists who had the technical know-how to use them. In recent years, generative art tools have evolved rapidly, expanding the potential for AI-driven creativity and transforming how artists engage with digital media. DALL-E 2, a development of the original DALL-E model, was introduced by OpenAI in April of 2022 and further advanced language-guided image generation, allowing for the creation of high-quality images from detailed text prompts. DALL-E 2 represents a leap in the ability of AI to understand and interpret language in a nuanced way, translating complex descriptions into images that capture not only the content but also the tone and aesthetic implied by the prompt. This capability enables artists to bring conceptual ideas to life directly from written descriptions, supporting creativity in fields like advertising, concept art, and visual storytelling. The high level of fidelity and specificity in DALL-E 2's outputs illustrates the evolving sophistication of language-guided AI models, which continue to make creative production more accessible and collaborative (Ma et al., 2024).

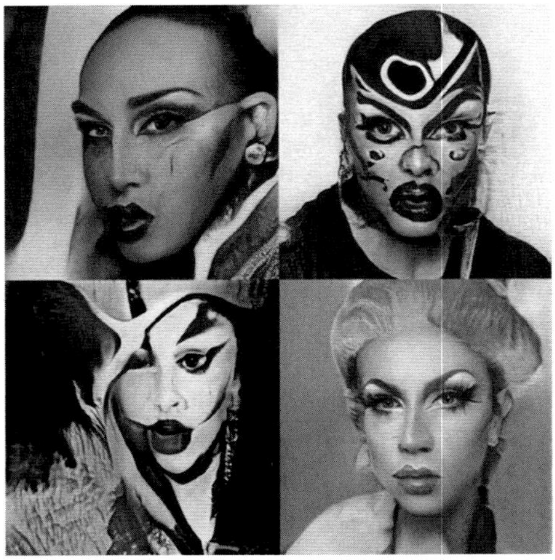

Figure 3.12 Jake Elwes, *Zizi - Queering the Dataset*, film still, 2019 (Creative Commons License 2.0).

Another notable advancement is Midjourney (Fig. 1.4), introduced in July of 2022, which focuses on style transfer to transform images and videos into various artistic styles. This tool led to an explosion of AI art social media groups as it allows users to reimagine visual content by applying stylistic elements from famous art movements, such as Impressionism, Cubism, or Futurism, or even from specific artists' techniques. Midjourney's style-transfer capability provides unprecedented versatility in visual media, enabling artists to experiment with different aesthetics and reinterpret existing works through unique stylistic lenses. Given that style transfer is accessible, Midjourney supports both creative experimentation and the development of new visual languages, which have applications ranging from film to digital art (Tsidylo and Sena, 2023).

Another significant breakthrough in generative AI for art is Stable Diffusion (Fig. 1.3), released in August of 2022, which emphasizes generating stable, controllable models for image synthesis. Stable Diffusion addresses one of the challenges in AI-generated art: achieving consistent results while maintaining creative flexibility. Unlike earlier models, Stable Diffusion allows users to input more precise controls over the image generation process, offering a stable base while enabling customization in areas such as color, form, and composition. This stability has proven invaluable for artists seeking to refine their works or produce images that meet specific visual criteria, making Stable Diffusion a critical tool in professional creative workflows. Its balance between control and creativity has broadened the scope of AI applications, allowing artists to explore ambitious projects with reliable, high-quality outcomes (Zhang et al., 2024).

Reacting to the concerns over copyright, Adobe Firefly, launched in March of 2023, introducing an interactive and dynamic approach to generative AI. Firefly empowers artists to apply real-time visual effects to images and videos, allowing for instant feedback and interactive experimentation. Unlike traditional static generative models, Firefly emphasizes adaptability and user interaction, giving artists a sense of immediacy and hands-on control over the generative process. This tool is particularly valuable in fields like digital media and animation, where dynamic effects can enhance visual storytelling and bring a new dimension to static images. Through the integration of real-time interactivity, Firefly underscores the potential for AI to become a collaborative tool in creative projects, enhancing productivity and allowing artists to iterate rapidly on their ideas (Angelova, 2024).

The success of these models paved the way for generative AI to expand into video, with text-to-video models emerging in late 2022. By then, companies like Meta and Google unveiled their own text-to-video prototypes, including Meta's Make-a-Video and Google's Imagen Video, which translated textual prompts into short video clips. These models represented an enormous leap in complexity, as they required AI to not only generate coherent images but also to produce frames that flowed together seamlessly over time. More recently, in October 2023, advancements such as Runway's Gen-2 model and Adobe's Firefly for video further pushed text-to-video technology into more practical and accessible territory, enabling higher-quality outputs with enhanced user control.

The latest advancements in text-to-video GAI models have continued to push the boundaries of generative AI, with several prominent models emerging in the past year. In June 2024, Runway introduced its Gen-3 Alpha model, followed by the release of Act-One in October. Known for its innovative generative tools, Runway's Gen-3 Alpha offers users a sophisticated platform for creating high-fidelity videos by specifying details like camera movements, lighting, and scene aesthetics, giving content creators greater control over the final product. Act-One builds on this functionality, refining Runway's video generation capabilities and making it easier for users to produce complex, dynamic scenes based on simple prompts, positioning Runway as a leader in creative video production.

OpenAI's Sora, previewed in February 2024, marks the first foray of the company into video generation, expanding its AI offerings beyond text and image generation. Although still awaiting a public launch, Sora allows users to create video content directly from text prompts, representing a significant step in OpenAI's generative AI capabilities for multimedia production. While in September 2024, Synthesia released Synthesia 2.0, a tool focused on digital avatars and video creation for professional content. Synthesia 2.0 allows users to create videos featuring lifelike avatars capable of speaking in multiple languages, making it especially useful for businesses needing professional, customizable video content. This update enhances Synthesia's capacity for quick, adaptable video production, meeting the demands of global audiences with localized content options.

These recent models illustrate the rapidly advancing capabilities of generative AI in video production, each with distinct functionalities tailored to diverse needs, from marketing to high-quality film content. Also, these latest iterations offer increased resolution, extended video length, camera control, and improved stylistic coherence, allowing for more polished, professional results suitable for a variety of applications. This evolution from text-to-image to text-to-video marks a significant expansion in of capabilities, as the technology becomes increasingly versatile, enabling creators to produce complex visual narratives from simple prompts and broadening the scope of the role these systems are playing in digital content creation.

The latest advancements in generative AI are reshaping modern cinema, fundamentally changing how films are created, edited, and produced. AI tools like RunwayML's Gen-2 and DeepBrain AI are increasingly integrated into filmmaking workflows, assisting with tasks from scriptwriting and scene generation to real-time editing. These tools allow directors to envision complex scenes without relying entirely on traditional VFX resources, making production more efficient and accessible. In visual effects (VFX) and post-production, AI now plays a significant role in de-aging actors, generating realistic background environments, and even creating entire characters, providing filmmakers with greater flexibility and control over the final output. By accelerating processes like rendering and scene design, AI is reducing costs and shortening timelines, giving rise to a more streamlined approach to visual storytelling in cinema.

AI is also transforming the very nature of storytelling, allowing for unprecedented levels of creativity and experimentation in plot and character development. The film *Zone Out* (2018) (https://youtu.be/vUgUeFu2Dcw) showcases this potential of in storytelling, marking an early exploration into machine-generated narratives. Created through a collaboration between filmmaker Oscar Sharp and AI researcher Ross Goodwin, the film employed a neural network named Benjamin, which was fed numerous scripts and tasked with autonomously generating unique plotlines, character interactions, and dialogue. This experimental approach resulted in unconventional narrative structures and interactions, offering a surreal, dream-like experience that defied traditional storytelling norms (Upadhyay, 2023). The unique dialogue and character development of the film reflect the inherent unpredictability of ML, creating a story that is both familiar and oddly alien, underscoring the capacity to introduce new, unexpected perspectives into cinema.

The production process for *Zone Out* involved feeding the AI an extensive dataset of scripts and story elements, allowing it to "learn" the patterns and structures of storytelling before creating its own unique narrative. This method highlights the potential to autonomously develop story arcs and character dynamics without direct human scripting, pushing the boundaries of how films are conceptualized and produced. As generative models continue to improve, tools like openFrameworks (2023) are providing artists with the means to incorporate algorithmic systems into filmmaking, blending AI-generated visuals with traditional cinematic techniques (Chibalashvili et al., 2023). These hybrid methods enable filmmakers to create novel aesthetic and thematic experiences that reflect both human creativity and the unique, non-linear logic of machine learning, offering new ways to explore complex emotional and philosophical themes.

The implications for future filmmaking are profound. As models become more advanced, we could see systems autonomously generating entire films, from plot and dialogue to visual style, with minimal human intervention. This level of autonomy could allow filmmakers to create dynamic, adaptive plots that respond to real-time audience input or adjust based on thematic direction, providing highly personalized storytelling experiences. The potential for customizable character arcs and interactive plots, tailored to viewer preferences, could redefine engagement in film, allowing audiences to experience stories that feel personally crafted. The same year found mainstream services already showcasing such interactivity.

Interactive and immersive experiences are on the rise, with platforms like Netflix's *Bandersnatch* (2018) allowing viewers to make narrative choices and shape the direction of the storyline. This interactive approach to storytelling reflects the growing role in creating personalized experiences, where audiences have control over the unfolding narrative (McSweeney and Joy, 2019). AI also powers advancements in interactive advertising, such as shoppable videos, which suggest products in real-time during playback, making advertising a more engaging and seamless part of the viewing experience (Khoi and Le, 2024). Additionally, AI facilitates the creation of 360-degree experiences, AR, and VR content, allowing users to explore and interact with content more fluidly (Kim et al., 2023). By enhancing interactivity, AI helps bridge the gap between passive viewing and active engagement, providing audiences with a more tailored, immersive experience that aligns with the demands of modern digital consumers.

In video marketing, AI is transforming the way brands connect with audiences by creating highly customized ad content. Shoppable videos, for instance, enable viewers to purchase items directly within the video, offering a streamlined shopping experience powered by AI recommendations. Moreover, AI-driven algorithms are used to generate personalized narratives for ads, customizing content based on viewer preferences and behavior, thus increasing engagement and relevance. This level of customization allows brands to create narratives that resonate with individual viewers, making each ad experience unique and enhancing the effectiveness of marketing campaigns. As AI-powered marketing tools continue to develop, brands will have even greater ability to reach targeted audiences in ways that feel authentic and personalized, shifting the advertising landscape toward a more consumer-centric approach (Dong et al., 2023).

It is clear that these technologies are having a profound impact on video art, professional videography, and cinema has opened new creative avenues while challenging traditional concepts of authorship, authenticity, and artistic value. This chapter traced the evolution of AI and creative machines from early experiments with automated drawing devices to the sophisticated generative models of today, highlighting the role of the technology in reshaping both artistic practices and commercial video production. AI now assists or even autonomously generates visual media, expanding what is possible within video creation and giving rise to a hybrid space where human creativity converges with machine-driven innovation.

Looking ahead, this generative potential to popularize video creation is vast. Autonomous filmmaking, where systems could generate entire films from script development to final production, presents exciting prospects for lowering barriers to entry in the industry. This is already empowering diverse voices, which will be discussed in the next chapter, and broadening the scope of perspectives in cinema, providing a more inclusive and accessible landscape for storytellers worldwide. At the same time, the ability to respond to audience feedback or integrate real-time interactions suggests a future where narrative and interactive film may blend, giving viewers more agency over their experiences and blurring the line between passive viewing and active engagement.

Yet, as AI becomes more embedded in visual media, critical questions arise about the nature and value of AI-generated content. The ease with which anyone can produce images or videos raises concerns about the sheer volume and perceived worth of AI-generated content. The reception of AI art has been mixed; while it has enabled content creation, some have questioned its artistic integrity and commercial viability. While examples such as the portrait of *Edmond de Belamy* receiving a bid of nearly half a million dollars, the limitations on copyrighting such imagery remain problematic for creatives. As well, uses in branding have mixed with examples such as the Glenlivet's use of AI art on a high-end whisky bottle highlighting the potential tension between AI-generated visuals and the artisanal traditions they are intended to represent (Foley, 2024). AI art also raises complex questions about whether AI creations can evoke the same sense of meaning and authenticity traditionally associated with human-made works.

As AI becomes a pervasive aspect of modern life, artists are likely to find new ways to engage with it, whether by using it as a tool, critiquing its effects on society, or exploring its limitations. Thought-provoking projects like Ai Weiwei's *81 Questions*, which featured a public conversation between the artist and ChatGPT, demonstrate how artists can utilize AI not just as a medium but as a subject of critique, reflecting on its influence and implications. The view of the artist that easily reproducible art has lost meaning highlights a growing sentiment: AI-generated content may become ubiquitous, but the intrinsic value of human creativity and intention will likely continue to hold a unique place within the art world (Gaskin, 2024).

The latest developments in AI, robotics, and art highlight this continuing fascination with how machines intersect with human creativity. Ai-Da, a humanoid AI robot developed by British gallerist Aidan Meller, exemplifies this merging of art and robotics. The auction debut of Ai-Da, with her painting *A.I. God. Portrait of Alan Turing* (2024), marks a significant milestone: it is the first time artwork from a humanoid robot will be sold at Sotheby's. The portrait of Alan Turing (1912–1954), a foundational figure in AI and computer science, is not only a tribute to the legacy of the English mathematician but also a commentary on the ethical complexities surrounding both human and machine intelligence. According to Meller, what sets Ai-Da's work apart from other AI-generated pieces is the physical presence of the artist herself—a robot that embodies AI as both creator and medium. Ai-Da's participation in the auction and the high valuation of her work, estimated between $120,000 and $180,000, reflect growing public interest in AI art and its implications for traditional artistic values. This integration of AI, art, and robotics introduces new questions about authorship, value, and the role of machines in creative expression (Porterfield, 2024).

As AI in visual media continues to advance, its role will remain complex—both a tool for creation and a subject for reflection, critique, and redefinition. The future of generative AI in video and art points toward a collaborative convergence where human creativity and machine intelligence interact dynamically. This evolving relationship promises to expand the scope of what can be imagined and expressed, making it an era of limitless potential yet demanding thoughtful engagement to preserve meaning, authenticity, and artistic integrity.

References

Aiello, L. C. (2016). The multifaceted impact of Ada Lovelace in the digital age. *Artificial Intelligence,* 235: 58–62.

Åkervall, L. (2015). McMansion of media excess: Ryan Trecartin's and Lizzie Fitch's SITE VISIT. NECSUS. *European Journal of Media Studies,* 4(1): 279–286.

Amyes, T. (2013). Video, film and pictures. pp. 58–67. *In*: *Audio Post Production for Television and Film*. Routledge.

An, D. (2022). Technology-driven virtual production: the advantages and new applications of game engines in the film industry. *Revista Famecos,* 29(1): e43370–e43370.

Angelova, N. (2024). The capabilities of the art-oriented artificial intelligence Adobe Firefly and its visual advantages and disadvantages. *Journal of Fundamental Sciences and Applications,* 30(1): 1–10.

Berio, D., Akten, M., Leymarie, F. F., Grierson, M. and Plamondon, R. (2017, June). Calligraphic stylisation learning with a physiologically plausible model of movement and recurrent neural networks. pp. 1–8. *In*: *Proceedings of the 4th International Conference on Movement Computing*.

Bertelsen, L. and Short, A. R. (2023, August). State of the Art: A review of AI art generation methods for rigorous design. *In*: *International Design Engineering Technical Conferences and Computers and Information in Engineering Conference* (Vol. 87318, p. V03BT03A066). American Society of Mechanical Engineers.

Bhardwaj, R., Kadam, T., Waghule, S., Shendurkar, S. and Sarag, B. (2023, August). Creative AI using DeepDream. pp. 281–296. *In*: *International Conference on Image Processing and Capsule Networks*. Singapore: Springer Nature Singapore.

Boucher, B. (2023). 6 artists who were using artificial intelligence before ChatGPT. *Art*. Retrieved: https://www.artsy.net/article/artsy-editorial-6-artists-artificial-intelligence-chatgpt.

Bregler, C. (2007). Motion Capture Technology for Entertainment [In the Spotlight]. *IEEE Signal Processing Magazine,* 24: 160–158. https://doi.org/10.1109/MSP.2007.906023.

Buchan, S. (2013). Mechanics and magic. pp. 23–69. *In*: *Pervasive Animation*. Routledge.

Carter, C. (2019). Hyper-realism in the Adventures of Tintin. *International Journal of Computer Graphics and Animation (IJCGA),* 9(4): 1–12.

Chibalashvili, A., Savchuk, I., Olianina, S., Shalinskyi, I. and Korenyuk, Y. (2023). Creative coding as a modern art tool. BRAIN. *Broad Research in Artificial Intelligence and Neuroscience,* 14(2): 115–127.

Chiglintsev, E. A. (2015). Reception of the Icarus myth in the mass art of the late 20th–21st century. *Terra Sebus. Acta Musei Sabesiensis. Special Issue 2014. Russian Studies. From the Early Middle Ages to the Present Day*, 177–186.

Cohen, H. (1995). The further exploits of AARON, painter. *Stanford Humanities Review,* 4(2): 141–158.

Cohen, P. (2016). Harold cohen and AARON. *AI Magazine,* 37(4): 63–66.

de Almeida, J. (2020). The dot and the line: Drawing amongst computers. *A Companion to Contemporary Drawing,* 407–429.

de Panafieu, C. W. (1984). Automata—a masculine utopia. pp. 127–145. *In*: *Nineteen Eighty-four: Science between Utopia and Dystopia*. Dordrecht: Springer Netherlands.

Deterding, S., Hook, J., Fiebrink, R., Gillies, M., Gow, J., Akten, M., ... & Compton, K. (2017, May). Mixed-initiative creative interfaces. pp. 628–635. *In*: *Proceedings of the 2017 CHI Conference Extended Abstracts on Human Factors in Computing Systems*.

Dong, X., Liu, H., Xi, N., Liao, J. and Yang, Z. (2023). Short video marketing: what, when and how short-branded videos facilitate consumer engagement. *Internet Research, (ahead-of-print).*

Draves, S. Electric Sheep (1999). *URL http://www. electricsheep. org.*

Figueira, A. and Vaz, B. (2022). Survey on synthetic data generation, evaluation methods and GANs. *Mathematics,* 10(15): 2733.

Filson, L. (2017). Magic and mechanics: the late-renaissance automata of Francesco I de'Medici. *In*: *Evidence in the Age of the New Sciences*. Springer.

Fitzgerald, W. (1984). Aeneas, Daedalus and the labyrinth. *Arethusa,* 17(1): 51–65.

Foley, J. (2024). This whisky maker's AI packaging design feels weirdly off-brand. *Creative Bloq.* February 22, 2024: https://www.creativebloq.com/news/the-glenlivet-ai-art-packaging-design.

Franco, F. (2022). Interview with Frieder Nake: interview recorded in Bremen, 7 March 2017, and Venice, 20 October 2017. pp. 101–125. *In: The Algorithmic Dimension: Five Artists in Conversation.* Cham: Springer International Publishing.

Frichot, H. (2015). Matthew Barney's cremaster cycle revisited: Towards post-human becomings of man. *Angelaki,* 20(1): 55–67.

Gan, C., Gan, Z., He, X., Gao, J. and Deng, L. (2017). Stylenet: Generating attractive visual captions with styles. pp. 3137–3146. *In*: *Proceedings of the IEEE Conference on Computer Vision and Pattern Recognition.*

Gao, Y. and Liu, H. (2023). Artificial intelligence-enabled personalization in interactive marketing: a customer journey perspective. *Journal of Research in Interactive Marketing,* 17(5): 663–680.

Gaskin, S. (2024). Ai Weiwei has 81 questions for artificial intelligence. *OCULA.* January 9, 2024: https://ocula.com/magazine/art-news/ai-weiwei-has-81-questions-for-ai/.

Gauthier, J. (2014). Conditional generative adversarial nets for convolutional face generation. *Class Project for Stanford CS231N: Convolutional Neural Networks for Visual Recognition, Winter Semester,* (5): 2.

Gerolemou, M. (2022). *Technical Automation in Classical Antiquity.* Bloomsbury Publishing.

Goenaga, M. A. (2020). A critique of contemporary artificial intelligence art: Who is Edmond de Belamy?. *AusArt,* 8(1): 49–64.

Goodfellow, I., Pouget-Abadie, J., Mirza, M., Xu, B., Warde-Farley, D., Ozair, S. et al. (2020). Generative adversarial networks. *Communications of the ACM,* 63(11): 139–144.

Gordon, D. (2003). *Play Dead: Real Time (February 22–March 29, 2003, Gagosian Gallery, New York).*

Greenhalgh, C. M. B. (2008). Chapter eight. King, Pope, Emir and Caliph: Europe and the islamic building boom. pp. 327–361. *In*: *Marble Past, Monumental Present.* Brill.

Hanhardt, J. G. (2006). Nam June Paik (1932–2006) Video Art Pioneer. *American Art,* 20(2): 148–153.

Heilo, O. (2022). The ʿAbbāsids and the Byzantine Empire. pp. 339–370. *In*: *Baghdād.* Brill.

Holzer, J. (2021). Covid-19/expose. *Critical Inquiry,* 47(S2): S121–S122.

Huang, Y., Lv, S., Tseng, K. K., Tseng, P. J., Xie, X. and Lin, R. F. Y. (2023). Recent advances in artificial intelligence for video production system. *Enterprise Information Systems,* 17(11): 2246188.

Iafrate, A. (2016). The Solomonic Throne in Constantinople. pp. 55–105. *In*: *The Wandering Throne of Solomon.* Brill.

Johnston, J. (2008). *The Allure of Machinic Life: Cybernetics, Artificial Life, and the New AI.* MIT Press.

Katz, J. H. and Ellis, L. M. (2024). Dances with avatar: how creators can reduce the novelty of their work to achieve more creative success. *Academy of Management Review,* (ja), amr-2022.

Khoi, N. H. and Le, A. N. H. (2024). Real-time interactivity and impulsive buying in livestreaming commerce: The focal intermediary role of inspiration. *International Journal of Human–Computer Interaction,* 40(11): 2938–2953.

Kim, J. H., Kim, M., Park, M. and Yoo, J. (2023). Immersive interactive technologies and virtual shopping experiences: Differences in consumer perceptions between augmented reality (AR) and virtual reality (VR). *Telematics and Informatics,* 77: 101936.

Krauss, P. (2024). What is Artificial Intelligence? pp. 107–112. *In*: *Artificial Intelligence and Brain Research: Neural Networks, Deep Learning and the Future of Cognition.* Berlin, Heidelberg: Springer Berlin Heidelberg.

LaRocco, M. (2018). *Video Camera Technology in the Digital Age: Industry Standards and the Culture of Videography* (Doctoral dissertation, University of Southern California).

Levinson, J. (2015). Time and Time Again: Temporality, Narrativity, and Spectatorship in Christian Marclay's" The Clock". *Cinema Journal,* 88–109.

Lipton, L. and Lipton, L. (2021). Early adopters: electronic cinematography and CGI. *The Cinema in Flux: The Evolution of Motion Picture Technology from the Magic Lantern to the Digital Era,* 691–699.

Loock, K. (2021). On the realist aesthetics of digital de-aging in contemporary Hollywood cinema. *Orbis Litterarum,* 76(4): 214–225.

Luger, M. (2015). Poetry as monument: Jenny Holzer and the memorial poems of 9/11. *Memory Studies,* 8(2): 183–196.

Ma, X., Tao, Z., Wang, Y., Yu, H. and Wang, Y. (2015). Long short-term memory neural network for traffic speed prediction using remote microwave sensor data. *Transportation Research Part C: Emerging Technologies,* 54: 187–197.

Ma, W. D. K., Lahiri, A., Lewis, J. P., Leung, T. and Kleijn, W. B. (2024, March). Directed diffusion: Direct control of object placement through attention guidance. pp. 4098–4106. *In*: *Proceedings of the AAAI Conference on Artificial Intelligence* (Vol. 38, No. 5).

Mazzone, M. and Elgammal, A. (2019, February). Art, creativity, and the potential of artificial intelligence. *In*: *Arts* (Vol. 8, No. 1, p. 26). MDPI.

McAlister, A. R., Alhabash, S. and Yang, J. (2024). Artificial intelligence and ChatGPT: Exploring Current and potential future roles in marketing education. *Journal of Marketing Communications,* 30(2): 166–187.

McCourt, F. (2012). An examination of the mechanisms of movement in Heron of Alexandria's on automaton-making. pp. 185–198. *In*: *Explorations in the History of Machines and Mechanisms: Proceedings of HMM2012*. Dordrecht: Springer Netherlands.

McSweeney, T. and Joy, S. (2019). Change your past, your present, your future? Interactive narratives and trauma in Bandersnatch (2018). *Through the Black Mirror: Deconstructing the Side Effects of the Digital Age,* 271–284.

Meigh-Andrews, C. (2013). *A History of Video Art.* A&C Black.

Mezei, P. (2021). From Leonardo to the Next Rembrandt–the need for AI-Pessimism in the age of algorithms. *UFITA Archiv für Medienrecht und Medienwissenschaft,* 84(2): 390–429.

Moran, M. E. (2011). The history of robotic surgery. pp. 3–24. *In*: *Robotics in Genitourinary Surgery.* London: Springer London.

Morigi, M. (2004). New Media, New Museum Practices? Museums Respond to the Challenges of Digital Art.

Moussa, I. and Fekry, W. (2022). The influential impact and contributions of the scientific heritage of Mouseion's scholars towards renaissance and present-day technologies. *Journal of Tourism, Hotels and Heritage,* 4(1): 55–78.

Neshat, S., Azari, S., Toloui, S. and Ferydoni, P. (2009). *Women without Men.* Artificial Eye.

Nordin, A. N. and Ramli, N. (2020). Regenerating muslim inventors: the present future. *Ulum Islamiyyah,* 31: 1–18.

Oliveira, J. (2021). Computer art in Portugal. *Art, Museums and Digital Cultures*, 94–106.

Paik, N. J. (1963). Exposition of Music–Electronic Television. 1963. *Leaflet Printed for the Show.*

Pallant, C. (2007). Tarantino the cartoonist. *Animation,* 2(2): 171–186.

Paul, C. (2023). *Digital Art.* Thames & Hudson.

Pickering, A. (2007). Ontological theatre Gordon Pask, cybernetics, and the arts. *Cybernetics & Human Knowing,* 14(4): 43–57.

Porterfield, C. (2024). Painting by AI robot Ai-Da could bring more than $120,000 at Sotheby's. *The Art Newspaper*. October 22, 2024: https://www.theartnewspaper.com/2024/10/22/sothebys-ai-da-robot-auction-alan-turing-portrait-artificial-intelligence.

Rayner, J. (2013). *The Cinema of Michael Mann: Vice and Vindication.* Columbia University Press.

Reddy, V. S., Kathiravan, M. and Reddy, V. L. (2024). Revolutionizing animation: unleashing the power of artificial intelligence for cutting-edge visual effects in films. *Soft Computing,* 28(1): 749–763.

Rehak, B. (2007). The migration of forms: bullet time as microgenre. *Film Criticism,* 32(1): 26–48.

Ren, R. (2023). Ahead of His Time: Leonardo da Vinci's Contributions to Engineering. *Journal of Education, Humanities and Social Sciences,* 21: 18–25.

Ricketts, J. R. (2014). Robert Rodriguez's Magical Corridos: The El Mariachi Series and Latinos on Film. *Post Script-Essays in Film and the Humanities,* 33(3): 96–121.

Robinson, J. (2015). On Steve McQueen's Giardini and the Follies of Nations. *Visual Culture in Britain,* 16(1): 86–102.

Sager, M. (2024). Watch the new Toys 'R' Us ad generated by AI. *Newsweek.* June 26, 2024: https://www.newsweek.com/toysrus-brand-film-ad-ai-generated-sona-1917645.

Schnugg, C. (2019). *Creating ArtScience Collaboration: Bringing Value to Organizations.* Springer.

Schöpf, C. (1991). Decision at Prix Ars Electronica'90. *Leonardo,* 24(5): 619–619.

Shay, D. and Duncan, J. (1993). *The Making of Jurassic Park.* (No Title).

Sklar, R. (2001). The Case of Harmony Korine. *The End of Cinema as we know it: American Film in the Nineties*, 261–268.

Sovhyra, T. (2021). Artificial Intelligence and Issue of Authorship and Uniqueness for Works of Art (Technological Research of the Next Rembrandt). *Kult. i mystetstvo u suchasnomu sviti,* 22: 156–163.

Springmann, M. J. (2020). The Schlüsselfeld ship model of 1503. *The Mariner's Mirror,* 106(4): 390–407.

Stephensen, J. L. (2019, June). Towards a Philosophy of Post-creative Practices?–Reading Obvious' "Portrait of Edmond de Belamy". *In*: *Politics of the Machine Beirut 2019*. BCS Learning & Development.

Sugiarto, E., Kurniawati, D. W., Febriani, M., Fiyanto, A. and Imawati, R. A. (2021, March). Computer-based art in folklore illustration: development of mixed media digital painting in education context. *In*: *IOP Conference Series: Materials Science and Engineering* (Vol. 1098, No. 3, p. 032017). IOP Publishing.

Tao, Z., Lin, T. E., Chen, X., Li, H., Wu, Y., Li, Y. et al. (2024). A survey on self-evolution of large language models. *arXiv preprint arXiv:2404.14387.*

Tsidylo, I. M. and Sena, C. E. (2023). Artificial intelligence as a methodological innovation in the training of future designers: Midjourney tools. *Information Technologies and Learning Tools,* 97(5): 203.

Trail, G. (2021). Constructing a case for the first truly global contemporary medium. *Art Monthly Australasia,* (328): 68–71.

Upadhyay, L. (2023). Demystifying text generation approaches. *London Journal of Research In Computer Science and Technology,* 23(1): 15–19.

Uzun, A. and Vatansever, F. (2008). Ismail al Jazari machines and new technologies. *Acta mechanica et automatica,* 2(3): 91–94.

Vasileiadou, S., Kalligeropoulos, D. and Karcanias, N. (2003). Systems, modelling and control in ancient greece: Part 1: Mythical automata. *Measurement and Control,* 36(3): 76–80.

Viola, B. (2013). The Quintet of The Astonished (2000). *HD Video. Courtesy of the Artist.*

Violand, H. E. (1990). *Jean Tinguely's Kinetic Art or a Myth of the Machine Age*. (Volumes I-III). New York University.

Voto, C. (2022). From archive to dataset. Visualizing the latency of facial big data. *PUNCTUM. International Journal of Semiotics,* 8(1): 47–62.

Wang, R. (2021). Computer-aided interaction of visual communication technology and art in new media scenes. *Computer-Aided Design and Applications,* 19(S3): 75–84.

Watlington, E. (2022). Pretty gross: dialectics of desire and disgust in the reception of Pipilotti Rist. *Camera Obscura,* 37(1): 91–113.

Wei, R., Garcia, C., El-Sayed, A., Peterson, V. and Mahmood, A. (2020). Variations in variational autoencoders—a comparative evaluation. *IEEE Access,* 8: 153651–153670.

Zhang, N. (2020). Application of computer graphics and image software in marine graphic design. *Journal of Coastal Research,* 106(SI): 600–604.

Zhang, H., Xu, T., Li, H., Zhang, S., Wang, X., Huang, X. et al. (2018). Stackgan++: Realistic image synthesis with stacked generative adversarial networks. *IEEE Transactions on Pattern Analysis and Machine Intelligence,* 41(8): 1947–1962.

Zhang, Z., Zhang, Q., Lin, H., Xing, W., Mo, J., Huang, S. et al. (2024). Towards Highly Realistic Artistic Style Transfer via Stable Diffusion with Step-aware and Layer-aware Prompt. *arXiv preprint arXiv:2404.11474.*

Chapter 4
Case Studies
Contemporary Video Artist-Filmmakers

This section examines the work of contemporary video artists and filmmakers who integrate AI and digital technology to expand the boundaries of visual storytelling. Focusing on case studies of pioneering creators, it explores how these artists use AI-driven tools to experiment with form, narrative, and interactivity, blending video art and filmmaking to produce innovative works. The artists highlighted engage with themes of identity, representation, and societal issues, positioning AI as a tool for both social critique and creative expression. These case studies illustrate how AI allows for new approaches to authorship and originality, raising questions about the ethical implications of machine-assisted creation. By analyzing projects ranging from interactive installations to algorithmically generated video, this section offers insights into how generative AI is reshaping contemporary visual media and how artists are using it to challenge and reflect on today's technological landscape. This investigation reveals how artist-filmmakers are defining a new era of creativity, using digital and AI-enhanced media to navigate the complex intersections of art and technology.

4.1 Overview of AI in Contemporary Video Production

In the past year, AI video technology has made remarkable strides, transforming the creative landscape across a wide range of media (Rodgers, 2024). From social media platforms like TikTok and Instagram Reels, where content creators like Jake Fellman (https://www.youtube.com/c/jakefellman) are using AI to produce quick, captivating animations, to high-profile collaborations in music videos—such as the use of AI visuals in concerts by David Guetta (https://www.youtube.com/watch?app=desktop&v=M6tLJTwcp1g) and generative music videos like that of Snoop Dogg—AI is reshaping how artists engage audiences (Jetha, 2024; Perrelli, 2024). The adoption of AI in filmmaking is equally impactful, as seen in TCL's upcoming release of five films produced through its Film Machine AI accelerator, and the work of independent filmmakers who utilize tools like Runway Gen-2 to create visually complex, immersive scenes with limited budgets (Lafayette, 2024). Meanwhile, AI-powered video installations by artists like Refik Anadol (1985–) transform art galleries into interactive spaces that react to viewer input, merging real-time data with intricate generative designs to create ever-evolving visual experiences (Jebb, 2024).

In advertising, AI is radically transforming how brands engage with consumers, creating immersive, personalized experiences that were previously unachievable. Samsung's Neon avatars, for example, simulate lifelike interactions, teaching users skills like yoga and language with real-time responses (Langely, 2020). Similarly, Gucci uses Genies avatars to connect with younger audiences, allowing users to style their avatars in Gucci's digital collections, fostering long-term loyalty through interactive "digital closets" (Prance-Miles, 2020). In a unique approach, IBM offers emotionally responsive avatars like Sam and Sarah, designed to address environmental awareness and enhance workplace efficiency by assisting with tasks like room bookings (Kart, 2020).

Luxury and retail brands have also harnessed AI avatars to reimagine customer experience. Calvin Klein generated buzz with a campaign featuring virtual influencer Lil Miquela alongside Bella Hadid, targeting inclusivity and digital engagement (Walk-Morris, 2024). Yoox has launched YOOXMIRROR with a Daisy avatar, part of its virtual fitting room, lets shoppers try on digital outfits, blending fashion with interactive technology (Chitrakorn, 2021). Banking and automotive sectors have followed suit, with Karnataka Bank's DhIRA and Maruti Suzuki's Dave offering personalized service through digital avatars that answer customer inquiries and provide product information (NS, 2024).

As AI avatars become more sophisticated, they are transforming how brands connect with consumers, merging advanced tech with tailored interaction to enhance engagement and loyalty. Platforms like HeyGen and DeepBrain AI are now at the forefront of this transformation, allowing brands to create lifelike avatars and interactive videos that blend entertainment and commerce. HeyGen (https://www.heygen.com/), for example, is widely used to produce customizable shoppable videos where AI-generated avatars interact directly with viewers, guiding them through products and features in a natural, conversational style. These avatars can even speak in multiple languages, making it easy for brands to reach a global audience with tailored, region-specific messages. By integrating clickable shopping options within the video, HeyGen enables seamless transitions from content to commerce, increasing viewer engagement and conversion rates by offering a highly interactive shopping experience (Evagelou et al., 2024).

DeepBrain AI (https://www.aistudios.com/), on the other hand, specializes in creating hyper-realistic digital avatars with advanced facial expressions and gestures, which brands use to deliver more emotionally engaging and memorable interactions. These avatars are popular in e-commerce and customer service applications, where they serve as digital brand representatives, answering questions, demonstrating product features, and assisting in purchase decisions in real time. DeepBrain's avatars are also used to create personalized messages for individual viewers, making the marketing experience feel more human and directly relevant to each customer's preferences (Kaur et al., 2024). The result is a media landscape where the lines between film, art, advertising, and interactive content are increasingly fluid, with AI avatars at the center of this new approach to consumer engagement (Durgade et al., 2024).

The rise of AI-generated shoppable videos exemplifies how the line between entertainment and commerce is blurring. In these videos, viewers can purchase products directly from interactive elements embedded in the content, creating a seamless, engaging shopping experience that combines storytelling with transactional functionality (Tang and Chen, 2024). For example, when watching a video on skincare, viewers can click on products used by the avatar and immediately be redirected to a purchase page. Moreover, platforms like YouTube and Instagram are increasingly integrating AI tools that allow brands to produce short-form content tailored to individual preferences. Algorithms analyze viewer behavior and adjust content in real time, ensuring that ads are both visually appealing and contextually relevant (Kim and Kim, 2024). This fluid integration of AI into visual media is pushing the creative industry toward an era where the distinctions between art, film, advertising, and interactive media are distinguishable only by their intent, forging a new, multifaceted approach to digital storytelling (Reddy et al., 2024).

AI-driven content has particularly flourished on social media, where platforms like TikTok, Instagram Reels, and X (formerly Twitter) have become essential hubs for AI-powered short-form videos. Creators such as Karen X. Cheng (https://www.youtube.com/channel/UC0nBp9LDpQMXEvYtvIsTV1w) use AI to produce quick, visually striking animations, enabling efficient production that captivates audiences with rich, memorable content. Similarly, music videos have emerged as fertile ground for AI experimentation, as seen in collaborations with artists like The Weekend, who has incorporated AI-generated visuals into his concert teasers (https://x.com/theweeknd/status/1826290730652086275) and digital releases, creating an immersive, otherworldly experience for fans (Frazen, 2024). Indie musicians like Washed Out have also embraced tools such as OpenAI's Sora to produce AI-driven video content directed by Paul Trillo as seen in *The Hardest Part* (2024) (https://www.youtube.com/watch?v=-Nb-M1GAOX8) that transforms their music into surreal, immersive digital landscapes. Events like the monthly Machine Cinema (https://machinecinema.ai/) hackathon illustrate the collaborative potential of AI, where teams use tools like MidJourney and Runway to create surreal music videos, such as a human-dog hybrid overlaid with electronic beats, further highlighting the ability of these evolving tools to facilitate storytelling that blends music with visual expression.

Meanwhile, independent filmmakers are increasingly adopting the same tools to push creative boundaries within budget constraints, with AI-driven films gaining attention at festivals like the Sci-Fi London Film Festival (Gemünden, 2024). Films such as the indie sci-fi thriller *The Terra Age* (2023) (https://www.youtube.com/watch?v=nWY1lDIeEtw) (Fig. 4.1) leverages Midjourney, Runway Gen-2 and Epidemic Sound to design futuristic cityscapes and digital characters, demonstrating how AI can help build immersive worlds without the need for extensive visual

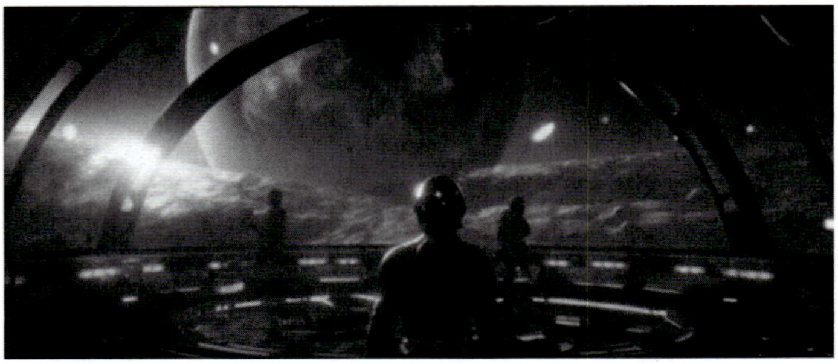

Figure 4.1 The Reel Robot, *The Terra Age*, film still, Runway ML, 2023 (Creative Commons License Zero).

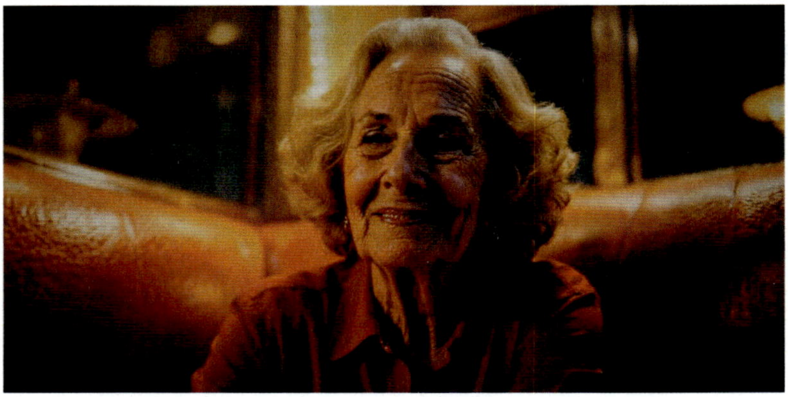

Figure 4.2 MetaPuppet, *Mnemonade*, film still, Kling, Runway ML, 2024 (Creative Commons License Zero).

effects teams. Other filmmakers are utilizing AI to enhance narrative elements as well. In *Mnemonade* (https://www.youtube.com/watch?v=_tIm8GBKRJ0&t=143s) (Fig. 4.2), a short sci-fi film created with RunwayML Gen-2, animates dream-like sequences, elevating the film's surreal aesthetic without requiring extensive VFX support. This award-winning, AI-powered short film tells the poignant story of an elderly woman with dementia who reconnects with lost memories through the flavors of forgotten recipes. Its emotionally powerful storytelling deeply resonated with audiences, earning the inaugural Culver Cup from AWS and FBRC.ai. Other projects, such as *Echoes of Tomorrow* (2024) (https://www.youtube.com/watch?v=yVF95ArZVOU) (Fig. 4.3) employ AI not only for visual effects but also for script generation, with AI co-authoring dialogue and plot elements to create narratives that blend human and machine inputs seamlessly. Such examples underscore the versatility of the tools to popularize high-quality cinematic production, making complex visuals, advanced effects, and innovative storytelling accessible to indie filmmakers (Izani et al., 2024).

Beyond traditional media, artists working in video installations are increasingly integrating AI to create immersive, interactive experiences. Notable figures already mentioned like Refik Anadol are transforming art installations into dynamic, evolving

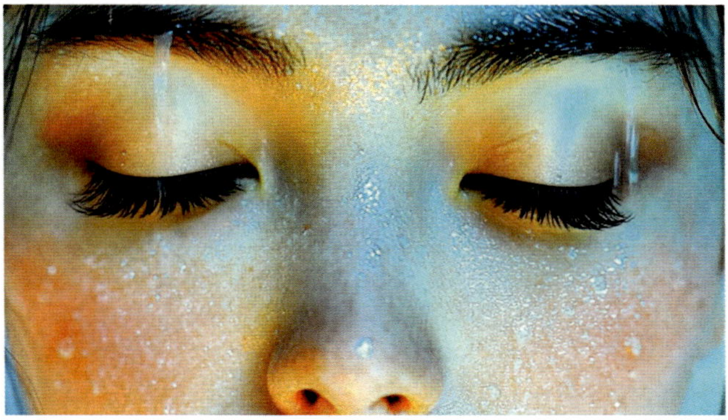

Figure 4.3 *Echoes of Tomorrow*, film still, 2024 (Creative Commons License Zero).

spaces by using vast datasets that respond to real-time viewer interactions. In his work *Machine Hallucinations* (2019), Anadol processed millions of architectural photographs to create a mesmerizing video installation that explores the concept of memory through machine intelligence, generating visuals that shift and change as the audience moves through the space (Cavdar, 2024). Similarly, Ian Cheng (1984–) is recognized for using AI to develop evolving, narrative-driven installations. In his *Emissaries* trilogy (2015–2017), Cheng created virtual worlds populated by AI-driven characters that interact with each other and adapt to environmental changes, blurring the line between art and video game. The work is constantly changing, providing viewers with a unique experience each time they encounter it, and exploring themes of chaos, adaptation, and the unpredictability of human experience (Kushnir, 2024).

Other artists, like Memo Akten (1975–), blend AI with custom algorithms to produce intricate, layered visuals that delve into the intersection of human expression and machine learning. The work of the artist *Learning to See: Gloomy Sunday* (2017) uses neural networks trained on the artist's own hand-drawn images to create visuals that question the nature of perception and emotion. The piece invites viewers to consider how machines interpret human emotions, challenging perceptions of both art and technology (McCormack and Grierson, 2024). Meanwhile, the Nigerian filmmaker Malik Afegbua creates AI-driven projects like his *The Elder Series* (2023) that disrupts stereotypes in AI-generated imagery by showcasing elderly African models in a high-fashion context, challenging conventional representations within AI art. Through training the model to generate dignified and stylish portrayals of elderly Black figures, Afegbua offers a striking critique of biases embedded in AI training data, showcasing how AI art can powerfully engage with cultural narratives and redefine visual norms (Dueno and Lopez-Figueroa, 2024). These artists illustrate how these new tools are pushing the boundaries of video installations, not only by enhancing viewer engagement but also by challenging and expanding cultural perceptions of technology, identity, and artistry.

Yet, while generative tools offer unprecedented freedom and flexibility, it also raises profound questions about the future of human creativity. Many artists and filmmakers grapple with an existential tension: AI can be a liberating tool that empowers new forms of expression, but it also poses a potential threat to the authenticity and individuality of creative work. As AI-generated content becomes more sophisticated and widespread, creatives face the paradox of a technology that both democratizes artistic production and challenges the fundamental value of human creativity, leaving them to navigate an evolving—and often ambivalent—relationship with the machines that now share their craft (Jääskeläinen et al., 2024).

As the following interviews with AI video artists reveal, the convergence of intelligent systems and creative media is profoundly reshaping the roles of artists and filmmakers. These pioneering creators, each bringing curiosity and an openness to experimentation, are navigating a landscape where AI is not only a tool but an integral part of their creative process. What unites these artists is a shared optimism about the potential of the technology to empower a diverse array of creators, regardless of background or location, to realize their own films. They emphasize the importance of daily experimentation, encouraging future creators to dive in and engage actively with these tools. One of the most exciting prospects is the potential for individual artists to produce feature-length films autonomously, a vision these creators believe is within reach thanks to the rapid advancements in generative image, video, sound, and VFX tools used as part of their workflow. Moreover, many of these artists are based outside traditional film industry hubs, signaling a decentralization of creative production that could redefine the global cinematic landscape, expanding the range of voices, styles, and narratives shaping the future of film. Through these interviews, it becomes clear that AI is not only transforming how we create art but also broadening the scope of who can participate in the creation of cinematic experiences.

4.2 Interviews with Leading AI Video Artists

4.2.1 Seth Steinacher (United States): Embracing AI for Creative Freedom in Video Editing

This section features an in-depth conversation with Seth Steinacher (https://www.linkedin.com/in/sethsteinacher/), a dedicated early-stage career video editor from Waverly, Illinois, whose journey into AI reflects both the potential and complexities of this technology in creative fields (Hutson, 2024a). The unique background of Steinacher—growing up in a small town with limited resources and learning video production through self-driven experimentation—offers insight into the personal and professional evolution of an artist who has embraced generative tools as part of his workflow. His story illustrates the dual nature of AI in video, serving as both an invaluable tool and a challenging shift in the traditional approach to media. As he states, "AI tools are just like any other—they do the data heavy-lifting, but creativity remains ours to shape." Through this interview, Steinacher provides a nuanced view on how AI enhances the video editing process while preserving the human element essential to authentic storytelling.

The journey of the creator into video production began at a young age, as he experimented with early camcorder technology and basic editing tools. Growing up in Waverly—a small, central Illinois town with a population of only around 1,200—he lacked access to formal creative resources, leading him to adopt a self-taught approach to video editing. "I was part of the first generation to see

YouTube become a thing… I wanted to be part of that new world," Steinacher reflects, capturing the enthusiasm that drove him to explore digital media despite these limitations. Using tools like Windows Movie Maker (https://apps.microsoft.com/detail/9mvfq4lmz6c9?hl=en-US&gl=US), he recalls, was "bare bones as you can get and crashes every two seconds," forcing him to learn the value of adaptability and perseverance early on. His foundational experience in self-directed learning, a necessity given his homeschooled background, would later become instrumental in his transition to AI-driven media.

For Steinacher, video editing was not only a skill but also a means of expression that allowed him to navigate personal and cultural challenges. As he describes, "Trial and error was my teacher—I wasn't trained, so everything came from testing and figuring things out on my own." This process of experimentation shaped his perspective on technology and creativity, fostering a resilient mindset that would prove valuable as he incorporated AI tools into his work. Reflecting on his journey, Steinacher considers his early experiences as a testament to the power of self-guided education in developing technical and creative skills in challenging environments.

The introduction of developing generative tools into the workflow of the videographer marked a significant turning point in his career, offering both new opportunities and efficiency. Initially skeptical about the abilities of such systems to meet his creative standards, he was quickly impressed by tools like Opus (https://www.opus.pro), an automated clip generator, and Adobe Audition (https://www.adobe.com), which streamlined complex audio editing tasks. These technologies, he notes, allowed him to "generate clips in 30 minutes without breaking a sweat"—a feat previously unthinkable in traditional video production. Adobe Audition, for example, "literally could take this conversation and make it sound like a professional recording," highlighting how intelligent systems enabled him to improve production quality without formal training in audio engineering.

Steinacher explains that the tools helped him overcome specific technical challenges, allowing him to focus on creative direction rather than technical tasks. "I'm not an audio engineer," he says, "but AI lets me do things I wouldn't have been able to figure out on my own." This shift allowed the early-stage creator to produce high-quality content more quickly, ultimately transforming his approach to video editing. With AI handling much of the "data heavy-lifting," Steinacher found himself free to explore new creative ideas, underscoring the role of AI in enhancing productivity and artistic experimentation within the constraints of a fast-paced industry. Outside of creative industries, professionals across various fields have similarly recognized the value of offloading technical or repetitive tasks to intelligent systems, allowing them to focus on innovative solutions, strategic planning, and a broader, more holistic view of their work (Faulconbridge et al., 2023; Susskind and Susskind, 2022).

As Steinacher integrated AI tools into his work, he discovered how they unlocked new possibilities for creative freedom and efficiency. With Opus and Adobe Audition managing routine editing tasks, he could shift his attention to higher-level aspects of storytelling and visual composition. This change in workflow not only enabled him to handle more projects simultaneously but also opened avenues for experimentation that were previously impractical. "I was struggling to keep up with the fast pace, and now I can focus on the fun part—the storytelling," Steinacher explains, capturing the relief and excitement that the technology brought to his creative process.

The capacity of the tools to enhance productivity has redefined how this creative approaches his projects, making what was once complex more accessible. Reflecting on the transformation AI brought to his work, he notes, "These tools give me creative freedom while handling the parts I don't enjoy as much." Through the reduction of the technical burdens of video editing, the tools have allowed him to reimagine his workflow, fostering a deeper engagement with storytelling and innovation. The experience here demonstrates how generative systems can serve as an invaluable ally in creative fields, facilitating a balance between efficiency and artistic integrity.

Despite his enthusiasm, Steinacher maintains a cautious perspective on the role of the technology in video production, acknowledging the potential for job displacement and over-reliance on automation. He regards AI as a powerful tool, but one that should complement rather than replace human creativity. "AI tools do the data heavy-lifting, but creativity remains ours to shape," he asserts, emphasizing his belief that technology cannot substitute for the intuition and artistic judgment unique to human creators. This sentiment reflects his view that while AI offers numerous benefits, it is essential to use these tools thoughtfully to maintain the integrity of creative work. The sentiment has echoed across studies of the impact the technology is already having on creative industries and the human-centered approach espoused by artists and designers (Wingström et al., 2024; Zhou and Lee, 2024). The perspective here is a nuanced one, recognizing both the opportunities this evolving technology presents and the ethical considerations that it raises. The creative believes that these generative tools can make video editing more accessible by making the process more accessible, particularly for creators who may lack extensive technical skills. Yet, he stresses the need for caution, particularly as the technology evolves to take on more sophisticated roles. "There's a balance to be struck," he says, underscoring his commitment to preserving the human element in art while embracing the advantages of AI.

Steinacher offers practical advice for those interested in entering the field, suggesting that newcomers incorporate generative production into their creative toolkit with an open mind and a willingness to experiment. He views AI as a powerful aid rather than a replacement for traditional skills, urging beginners to "dive in with curiosity and patience," as learning to use machine solutions requires exploration and adaptation. Starting with user-friendly tools will ensure aspiring video artists can gain hands-on experience and understand these new capabilities without feeling overwhelmed. The journey of this creative illustrates how AI-assisted platforms can lower entry barriers, equipping new artists with resources and possibilities that were once out of reach. With the right approach, Steinacher believes, AI can popularize creative expression and empower more people to tell stories through video. His encouragement reflects his own experience with AI, which has expanded his creative horizons and reinforced the importance of persistence and adaptability.

Looking forward, Steinacher envisions a future where AI will enable unprecedented levels of creative independence, allowing individual artists to produce complex video projects from start to finish. He imagines a fully integrated suite with tools capable of handling every stage of production, from scripting and editing to post-production effects. "I dream of a time when one person can create an entire film, from writing to post-production, through AI alone," he says, capturing his excitement for the possibilities of generative technology. His vision reflects a belief that continues to evolve, broadening the scope of what artists can achieve independently.

The aspirations of the creator for such tools extend beyond efficiency; he sees it as a vehicle for greater innovation and inclusivity in video production. With the anticipated release of multimodal tools like ChatGPT 5, which promises capabilities in text, audio, and video, he believes the technology will open new pathways for storytelling and artistic experimentation. "This is a new frontier in storytelling," he asserts, underscoring his enthusiasm for a future where creators of all backgrounds can explore and innovate without traditional limitations. For Steinacher, AI represents not just a tool but a transformative shift in the creative landscape, one that holds the promise of redefining both the production process and the possibilities available to artists worldwide. As we move to the next interviews, the same sentiments are expressed and become a trend regardless of the video artist's technical or cultural background.

4.2.2 Stephen Erin Dinehart IV (United States): Advancing Co-Creative Storytelling with AI in Interactive Media

In this segment, Prof. Stephen E. Dinehart IV (https://narrativedesigner.com/) Assistant Professor of Film, Animation & New Media at the University of Tampa, Florida, shares insights into his journey and perspectives on AI in video production (Hutson, 2024b). Known for pioneering narrative design in the gaming industry, the distinguished career of Prof. Dinehart spans over two decades, marked by collaborations with major entities such as Nintendo, Warner Bros., and Universal Studios. His recent achievement, *Mario Kart: Koopa's Challenge* (https://www.youtube.com/watch?v=DKDpWMrZSJA), received the prestigious Golden Ticket Award for Best New Family Attraction in 2023, underscoring his innovative influence on family entertainment. Dinehart views the impact of AI on creativity and production as a natural extension of his dedication to interactive media and transdisciplinary creativity, seeing it as a tool that enhances storytelling potential and expands the boundaries of audience engagement.

The career of this educator-artist in video and interactive media began in the world of AAA video games, where he crafted narrative frameworks that added depth to gameplay and heightened player immersion. He describes his approach as "building worlds that respond and adapt," a perspective that has informed his work across multiple platforms, from video games like *Constantine* and *Batman Begins* to theme park attractions like *Koopa's Challenge*. His early work was characterized by a desire to intertwine technology and storytelling, setting the stage for his seamless transition into the AI-driven media landscape. His background in narrative architecture has made him a thought leader in using interactive storytelling techniques to deepen audience engagement—a foundation that now informs his use of new tools.

Building on his narrative expertise in interactive media, Dinehart has recently ventured into AI filmmaking with *Trojan Horse* (2023) (https://www.youtube.com/watch?v=RFqQNrSB1TE) (Fig. 4.4), a short film created in 4 K for the Sci-Fi London 48 hr Film Challenge. The film explores themes of trust, deception, and technological control within a futuristic landscape where artificial intelligence and human autonomy intersect in complex, ethically ambiguous ways. Through this story, Dinehart prompts viewers to consider the implications of AI on human agency, a theme consistent with his larger body of work. The tight narrative structure and compelling subject reflect his expertise in building immersive worlds that resonate emotionally and intellectually. The film also exemplifies the approach of the artist to these intelligent systems as tools for enhancing traditional cinematic techniques. Through such utilization for rapid prototyping and execution, he achieves high visual clarity and depth within a limited timeframe, allowing him to focus on narrative intricacies, which reflects his philosophy of "co-creation," where AI complements human creativity rather than replacing it.

With AI handling technical tasks, Dinehart emphasizes storytelling and thematic development, demonstrating the demonstrable potential of generative tools as a collaborative partner in delivering impactful, adaptive visual narratives. The theme runs through his career and echoes the beliefs of those interviewed across the globe. Reflecting on his career, Dinehart emphasizes the importance of integrating storytelling with technological innovation. "Story is what keeps people invested," he notes, underscoring his belief that narrative is central to meaningful engagement, whether in video games or AI-enhanced video production. His commitment to storytelling as a vehicle for immersive experiences has been a guiding principle throughout his professional journey. This focus on narrative has shaped his approach to AI, allowing him to apply his expertise in creating compelling, user-driven content across a variety of media.

With the introduction of these new tools into his workflow, Dinehart has been able to enhance his creative process, leveraging the capabilities to streamline technical tasks and augment storytelling potential. He discusses how generative tools enable him to quickly prototype narrative scenes, a function that he finds invaluable for developing complex, adaptive storylines. "AI lets me build narratives that evolve as the viewer interacts," he explains, highlighting the ability of the technology to facilitate dynamic storytelling. The AI-driven approach aligns with his philosophy of "co-creation," in which technology and human creativity collaborate to produce richer, more engaging experiences. Dinehart sees AI as a tool that both amplifies and accelerates the creative process, offering possibilities that were previously unattainable in conventional video production. Through AI, he is able to experiment with visual effects, adjust story pacing, and refine character interactions—all while focusing on the broader creative vision. He observes, "AI doesn't replace the artist; it enhances the tools we have to bring stories to life," a sentiment that captures his balanced perspective on the role of AI in creative work. Through the automation of repetitive tasks, these systems enable Dinehart to channel his energy into storytelling innovation, exemplifying how technology can be harnessed to serve artistic goals.

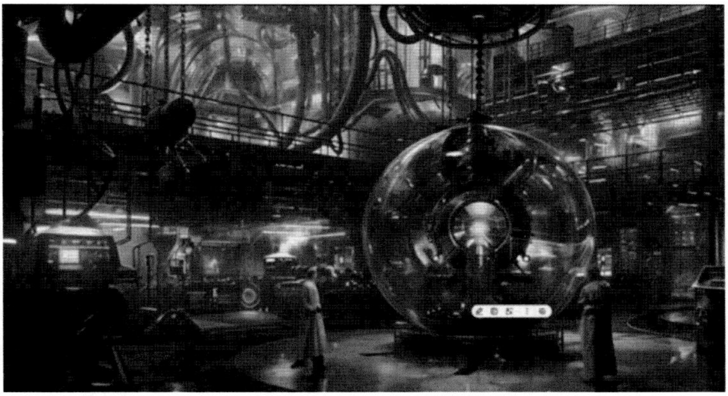

Figure 4.4 Stephen Dinehart, *Trojan Horse*, film still, 2023. Courtesy of Stephen Dinehart.

His adoption of generative tools is tempered by a critical awareness of its ethical implications, especially in a field where technology has the potential to alter audience perceptions and experiences. He acknowledges that while AI offers unprecedented creative freedom, it also raises questions about authenticity and artistic control. "We need to remember that AI is a tool, not a replacement for human insight," he asserts, stressing the importance of maintaining the human perspective within AI-driven projects. This stance reflects his belief in AI as a collaborative partner rather than an autonomous creator, emphasizing that technology should serve as an extension of the artist's vision, not a substitute.

The ethical considerations surrounding these systems in storytelling are particularly pertinent to Dinehart, given his background in creating immersive worlds where audience engagement is deeply personal. He expresses concern about the potential for generative technologies to be used in ways that prioritize profit over meaningful connection, cautioning against a "one-size-fits-all" approach to AI in media. In his teaching, he fosters a dialogue with students about responsible use, as he seeks to encourage a balanced approach to technology that respects both the creative process and the viewer's experience. His reflections highlight the nuanced relationship between innovation and integrity in the evolving landscape of media production.

For aspiring creators looking to integrate such tools into their video projects, Dinehart offers practical advice grounded in his own experiences with technology and storytelling. He emphasizes the importance of understanding AI as a complement to traditional skills, encouraging newcomers to explore its potential without losing sight of foundational techniques. "Start by mastering the basics—storytelling, pacing, character development—then use AI to expand those skills," he advises, underlining the value of a strong narrative framework in any AI-enhanced project. His guidance reflects a philosophy of "grounded innovation," where technology enhances rather than overshadows the core elements of storytelling.

The approach that the professor takes to the technology reflects a commitment to fostering creativity that is both accessible and impactful. He believes that such tools can lower the barrier to entry with the creative process, allowing new voices to participate in media production. Through the encouragement of emerging creators to experiment with generative tools within a structured narrative framework, he aims to empower a new generation of artists to harness technology thoughtfully. His advice serves as a reminder that while AI can enhance storytelling, the essence of a compelling narrative lies in the ability of the artist to connect with the audience on a human level.

Looking to the future, he is excited about the potential to further the concept of "co-creation," where technology and human ingenuity work in tandem to produce immersive, adaptive experiences. He imagines a media landscape where tools enable artists to create interactive stories that respond dynamically to audience input, a vision that aligns with his pioneering work in narrative design. "The future of storytelling is co-creative," he says, envisioning a world where AI supports rather than dictates the creative process. This model of co-creation not only enhances artistic flexibility but also invites audiences to become active participants in the narrative, transforming passive viewing into a collaborative experience.

This vision for the future of AI and creative industries reflects his lifelong commitment to pushing the boundaries of technology and creativity. With new advancements on the horizon, such as multimodal capabilities of handling both visual and narrative elements, he believes that the possibilities for interactive storytelling are endless. He sees these new tools as a catalyst for expanding the reach and impact of media, allowing artists to create stories that are both personal and universal. For Dinehart, AI represents a transformative force that, when used responsibly, has the potential to redefine the way stories are told and experienced, bridging the gap between creators and audiences in unprecedented ways.

4.2.3 Anup Gosavi (India/United States): Redefining Video as Data for Interactive, Adaptive Media

In this interview, Anup Gosavi (https://x.com/iamanupgosavi), co-founder of the video editing platform Spext (https://www.spext.co/), a Media AI company based out of San Fransisco, CA and Bangalore, India, explores the transformative potential of treating video as data rather than static files (Hutson, 2024c). Unlike traditional AI artists, the background of Gosavi in video editing and cloud technology has led him to innovative insights about the structural shifts needed to make interactive, customizable video experiences feasible. Originally from India and now based in the United States, Gosavi brings a multifaceted perspective to media, rooted in both technical development and practical applications. His journey reflects a drive to simplify video editing for informational content, noting that tools designed for cinematic production, such as Adobe Premiere Pro, often overcomplicate straightforward editing tasks. This realization led to the founding of Spext in 2019, which offers an intuitive, transcript-based editing system for webinars and lectures, making video more accessible for corporate communications.

The early experiences of the founder, including editing videos for weddings and events, illuminated the limitations of existing tools for informational video content. Observing that "movie-making tools were too complex for basic informational edits," he envisioned a simpler approach, one that would make video editing accessible for professionals needing quick, accurate adjustments. His breakthrough came with the rise of affordable, highly accurate speech-to-text technology, which allowed for a revolutionary editing method: using auto-generated transcripts where modifying text directly adjusted the corresponding video. "Imagine editing a video as easily as you edit a document," Gosavi remarks, describing the core innovation of Spext. This system, initially used within large enterprises for internal meetings and webinars, provides an efficient solution for summarizing and editing video content, akin to tools like Otter.ai (https://otter.ai/) but with integrated video functionality.

As his work progressed, the entrepreneur and his team expanded their platform into cloud infrastructure, enabling VideoDB (https://docs.videodb.io/)—a scalable solution to support both generated and recorded video. The shift from a singular editing tool to a cloud-based infrastructure reflects the vision of a more dynamic and responsive media landscape, one that facilitates instant video customization and real-time interaction. By focusing on "video-as-data," VideoDB aims to revolutionize how video is streamed, searched, and customized, reducing costs by up to 70% for large-scale applications. Gosavi views this approach as essential for transforming video into a truly interactive medium, moving beyond passive viewing into real-time, user-driven experiences.

A central tenet of his work is the concept of video as modifiable data rather than fixed files, a shift he sees as foundational to achieving interactivity in media. He argues that current file formats like MP4 are "designed for passive viewing," which inherently limits the potential for viewer interaction. In contrast, approaching video as data—socially shareable, searchable, and customizable—enables more flexible, scalable applications. Gosavi envisions a world where platforms like Netflix could offer not just a few alternate endings, as seen in *Bandersnatch* (2018), but potentially millions. "Why only 16 endings when you could have 16 million?" he asks, underscoring his vision for a future where video content adapts fluidly to user input and preference. This transformation, he believes, will make it feasible to generate vast quantities of video content tailored to individual viewers, breaking free from the limitations of traditional streaming.

This insight into video-as-data reflects his commitment to reimagining the media landscape in ways that prioritize adaptability and interactivity. Through the framing of video as a flexible medium, Gosavi suggests that AI could open the door to an era of hyper-personalized content, where individual viewers control not only what they see but how they engage with it. His concept extends beyond merely reducing production costs; it proposes a fundamental rethinking of video as an interactive and adaptive medium, paving the way for applications in entertainment, education, and beyond.

Despite the exciting possibilities of interactive video, Gosavi remains attuned to the ethical and legal complexities associated with this technology. He emphasizes the need for clear guidelines around copyright, particularly as AI-generated content often blurs the lines between original and derivative works. Gosavi contends that creators "should have the right to retain credit for their original work and to opt out of AI training systems." In his view, AI-generated content should carry disclosures, such as watermarks, to ensure transparency and respect for intellectual property. This stance is driven by a concern for maintaining ethical standards within an industry that, as Gosavi notes, has yet to fully address the ramifications of AI on creative ownership.

In addition to copyright, Gosavi highlights the importance of brand consistency for companies using AI-driven content. Professional use cases, he argues, must prioritize maintaining a unique voice and style, ensuring that AI-generated material does not infringe on the creative expressions of others. He describes the current landscape as a "phase of new technology without clear ground rules," likening it to early internet browsers that lacked definitive legal and ethical structures. These reflections illustrate the need for proactive dialogue and safeguards as AI continues to reshape media production and consumption.

For professionals interested in AI video, Gosavi recommends a strategic approach that balances innovation with practicality. He advises creators to begin by exploring the new capabilities in enhancing existing workflows rather than attempting to replace foundational skills. "Start with small, manageable projects that allow you to see AI's strengths and limitations firsthand," he suggests, emphasizing the value of incremental integration. Through testing the role that can be played by these systems in specific tasks, such as content moderation or video summarization, professionals can develop a deeper understanding of how to harness these tools effectively without overrelying on automation. Such advice reflects a cautious optimism about the new role AI is playing in media, advocating for a balanced approach that enhances creative output while preserving human agency. His emphasis on targeted experimentation provides a roadmap for professionals navigating the rapidly evolving landscape of AI-driven video. By taking measured steps, he believes that creators can maximize AI's potential while mitigating the risks associated with technology that is still in its formative stages.

Looking ahead, Gosavi envisions a future where AI enables media to be more interactive and community-driven, blurring the lines between creator and audience. He imagines a world where viewers actively shape storylines, contributing to narratives in real-time much like the *Dungeons & Dragons* community co-creates unique experiences. In this vision, content creation would become a collective endeavor, with AI facilitating an unprecedented level of viewer engagement and participation. "Think of a Netflix episode with 16 million possible endings, all shaped by the audience," he speculates, pointing to a future where stories are not only told but also lived by the viewers themselves.

Beyond video content, Gosavi foresees AI enabling hyper-personalized media experiences that adapt to individual preferences, even speculating on brain-computer interfaces that could allow users to "think about what they want to watch." This integration of AI and personalized media holds transformative potential, not only for entertainment but also for targeted advertising and other commercial applications. Gosavi's vision underscores the limitless possibilities of AI-driven media, where technology and creativity converge to produce interactive, adaptive experiences that redefine the role of the audience in the storytelling process.

4.2.4 PJ Accetturo (United States): Democratizing AI-Driven Storytelling and Interactive Media

In this section, PJ Accetturo (https://x.com/PJaccetturo/status/1842266006724759904), an award winning filmmaker and Hollywood insider, shares insights into his innovative approach to AI-integrated storytelling (Hutson, 2024d). As CEO of FilmPort.ai and Battle Island, Accetturo stands at the forefront of AI-driven video production, pushing the boundaries of traditional filmmaking and democratizing access to advanced video tools. His current project, *Ghosts of Ruin* (2024) (https://www.youtube.com/watch?v=XqM0VZnrvVI), a dystopian anime series exploring the future of gaming and virtual reality, exemplifies his commitment to exploring the intersection of AI and storytelling. Throughout his career, Accetturo has seen the potential for AI to transform the creative process, asserting that "the future of narrative filmmaking won't be constrained by Hollywood's traditional model but will instead empower individual creators." This vision encapsulates his belief that generative AI has the power to reshape entertainment, opening new avenues for both large-scale productions and solo creators.

The career of Accetturo began with traditional filmmaking, where he gained recognition through high-profile projects for brands like the Atlanta Braves and National Geographic. With over fifteen years of experience in documentary and commercial work, he brings a deep understanding of the creative process to his AI-integrated ventures. "I grew up as a one-man army filmmaker," he explains, highlighting how his experience in single-handedly managing production helped him adapt quickly to new technologies. His journey into AI began as a pragmatic choice: with AI tools becoming increasingly accessible, he recognized the potential to streamline complex video production tasks. This transition allowed him to create highly detailed content faster and with fewer resources, empowering him to explore new narrative possibilities that align with his vision for more democratized storytelling.

Reflecting on the evolution of video technology, Accetturo draws parallels between the rise of DSLR cameras in 2004 and the advancements today in AI. He argues that, just as affordable digital cameras once disrupted traditional filmmaking, AI will disrupt narrative production by making advanced tools accessible to creators worldwide. "The same way YouTube and Instagram democratized content creation, AI will do the same for narrative filmmaking," he asserts, emphasizing the significant potential of AI for both amateurs and seasoned professionals. This shift, he believes, will challenge the traditional Hollywood model, allowing creators outside the studio system to produce high-quality films and series independently.

As CEO of FilmPort.ai (https://www.filmport.ai/), Accetturo has integrated generative tools to streamline and scale production, particularly through tools that reduce the time and cost associated with rendering and editing. His team utilizes generative models to expedite processes like character design, background generation, and scene consistency. "AI lets us build worlds faster, with more creative control," he explains, noting that tools like MidJourney and NVIDIA's generative platforms allow his team to experiment with visuals and iterate quickly. Despite the current limitations of the technology in feature-length films, Accetturo advocates for episodic formats that better suit the current capabilities of generative tools, suggesting that "weekly series with Patreon-like subscriptions" represent a sustainable model for independent creators. By optimizing AI for shorter, serialized content, he believes creators can maintain both quality and viewer engagement.

This episodic approach aligns with his philosophy that AI should enhance—not replace—traditional storytelling. He observes that the strength of the tools lie in their ability to handle repetitive or time-consuming tasks, allowing artists to focus on narrative depth and creativity. "AI doesn't write the story for you, but it can handle the data-heavy lifting," he notes, pointing to its role in reducing the burden of labor-intensive tasks. This emphasis on AI as an enabler of creativity underscores Accetturo's commitment to a collaborative model, where technology supports artistic vision rather than dictating it.

His work with generative AI also brings to light the complex ethical considerations surrounding intellectual property in video production. With AI models like MidJourney often training on unlicensed data, the issue of copyright has become a pressing concern. "AI-created content needs transparency, especially in commercial contexts," he argues, highlighting the need for tools that verify the originality of AI outputs. Accetturo's team is developing a "copyright checker" to analyze AI-generated visuals against existing databases, allowing creators to detect similarities with copyrighted materials. He describes this tool as a necessary step to "reduce liability and maintain ethical standards," particularly as AI adoption accelerates in advertising and media.

Beyond copyright, Accetturo acknowledges the broader ethical challenges posed by AI, from maintaining brand consistency to preserving creative authenticity. He expresses frustration with ethical but visually inferior models like Adobe Firefly, which rely solely on licensed data but often lack the visual sophistication of models with broader datasets. This tension, he suggests, reflects an ongoing challenge for AI artists, who must balance ethical concerns with the practical need for high-quality output. "We're in a transition phase," he remarks, where AI's potential is vast but its ethical and legal frameworks remain underdeveloped. Accetturo's reflections underscore the need for responsible AI use, advocating for a future where creators can access powerful tools without compromising ethical integrity.

For aspiring filmmakers interested in AI, Accetturo offers a straightforward piece of advice: "Create every day." He encourages young creators to approach AI with a sense of exploration and consistency, stressing that "practice and curiosity are more important than perfection." Drawing on his own experiences, he believes that constant experimentation with AI tools allows creators to develop a unique style and build confidence in their work. Accetturo points to his own ritual of "daily creation," where he uses AI to generate new worlds and explore different artistic concepts, describing it as an essential exercise for creative growth. This iterative approach, he argues, will enable creators to navigate the rapidly evolving landscape of AI in media while cultivating their distinct voices.

This advice reflects a belief that foundational skills—like color theory, narrative structure, and visual composition—remain critical, even as AI automates many technical aspects of production. By mastering these fundamentals, he suggests, creators can better leverage AI tools to enhance, rather than overshadow, their artistic vision. His philosophy underscores the importance of building technical proficiency alongside creative intuition, empowering a new generation of artists to harness these new tools as a medium for personal and professional expression.

Looking to the future, Accetturo envisions a world where AI enables fully interactive and immersive storytelling experiences, allowing viewers to engage with narratives in unprecedented ways. Inspired by the potential of brain-computer interfaces and the concept of video-as-data, he imagines a future where audiences can "think about what they want to watch," creating hyper-personalized content that evolves in real-time. "We're heading toward a future where everyone can be their own writer and director," he states, describing a vision where AI democratizes not only production but also viewer engagement. This shift, he believes, will redefine the relationship between creators and audiences, fostering a media landscape where storytelling becomes a shared, participatory experience. The vision of Accetturo for AI-driven media reflects his commitment to advancing the art of storytelling through technological innovation. He sees a future where AI empowers creators of all backgrounds to produce high-quality content, transforming both the scale and accessibility of video production. With generative AI continuing to evolve, Accetturo remains optimistic that the tools will become more sophisticated and ethically grounded, paving the way for a media ecosystem where creativity, accessibility, and interactive engagement coalesce.

4.2.5 Zahir Kahn (Canada): Integrating AI for Efficiency and Innovation in Visual Effects

This segment introduces Zahir Kahn, the Toronto-based AI video artist with over a decade of experience in visual effects (VFX) for high-end feature films, episodic series, and animation (Hutson, 2024e). He also launched LoneLion (https://www.youtube.com/@LonelionZK) as the main studio for his AI short films. As a prominent digital matte painter, Zahir contributed to acclaimed feature films like *1917*, the 2019 Academy Award winner for Best Visual Effects. This background in matte painting and environment art has shaped his perspective on the role AI should play in VFX, where he sees the tools as a solution to streamline complex visual tasks. He believes that their real value lies in the ability to solve the technical, repetitive aspects of VFX, allowing artists to focus more on the creative elements. "Eighty percent of my work is problem-solving," he notes, emphasizing the potential of AI to reduce the labor-intensive workload of environment artists. The perspective highlights a pragmatic approach, where AI complements but does not replace human input, ultimately enhancing artistic efficiency and output quality.

The early career of Kahn was shaped by meticulous work in VFX, specializing in establishing shots, landscapes, and backgrounds for high-budget productions. This technical background provided him with a deep understanding of how to craft detailed visual environments—a skill that now informs his use of AI tools. "When I started, everything was manual; AI wasn't part of the equation," he recalls, underscoring the labor-intensive nature of his early work. Kahn's transition into AI tools stemmed from a desire to streamline his workflow, allowing him to focus more on creative composition rather than routine technical tasks. This shift aligns with his view that

AI should handle "data-heavy lifting," allowing artists to allocate their energy to the conceptual and aesthetic aspects of a project.

Reflecting on the demands of VFX work, Kahn identifies AI as a vital tool for reducing the time spent on repetitive, manual tasks. He describes his approach as a "hybrid model," blending AI, traditional VFX, and digital painting techniques to achieve the desired visual results. This model not only enhances efficiency but also supports artistic integrity, as it allows him to maintain control over critical creative decisions. For Kahn, AI represents an evolution in VFX, one that respects the fundamentals of traditional techniques while embracing the possibilities offered by technology.

An example of this can be seen in his recent 2024 release *The Siren 2: Nephilim's Mother* (https://www.youtube.com/watch?v=I5PsfQ_D31s) (Fig. 4.5) where Kahn showcases his innovative integration of AI and VFX to elevate the visual storytelling. The film begins with the haunting imagery of a shipwreck, setting a tense atmosphere as survivors cling to debris in the open ocean. Kahn leverages AI-driven VFX to enhance the naturalistic details of the water, debris, and surrounding environment, creating an immersive sense of isolation and vulnerability. As sharks circle and attack the survivors, the realistic textures and motion—enhanced by AI—intensify the suspense, making the ocean feel alive and perilous. The use of AI is particularly striking in the narrative shift, where a supernatural siren emerges from the depths. This character, rendered with detailed VFX and AI enhancements, captures an ethereal yet menacing beauty that contrasts with the earlier scenes of raw survival. Through the film, Kahn demonstrates how these tools can handle intricate background elements and supernatural details, allowing him to focus on narrative depth while pushing the boundaries of visual effects in a genre that demands both realism and fantasy.

At the same time, and while Kahn acknowledges the strengths of the new tools in generating visual elements, he is quick to address its current limitations. He argues that the role of AI in VFX is often misunderstood, especially by clients who assume that AI can fully replace traditional production techniques. "AI can help with filler shots or background landscapes, but it's not yet capable of producing

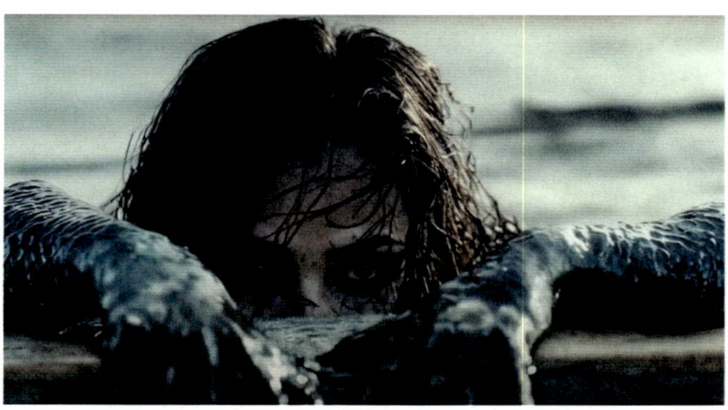

Figure 4.5 Zahir Kahn, *The Siren 2: Nephilim's Mother*, film still, 2024. Courtesy of Zahir Kahn.

core shots with the precision we need," he explains, pointing out that AI lacks the specificity and control required for high-stakes cinematic moments. For Kahn, the challenge is educating clients about these capabilities, ensuring that they understand when traditional methods are necessary for quality and authenticity. This pragmatic perspective on AI is informed by the experience of the filmmaker in a high-stakes industry, where visual fidelity and consistency are paramount. He notes that while AI can expedite certain aspects of production, its outputs still require refinement and integration with human-crafted elements. "You can't just take a client's brief and put it into text; someone has to make it work," he observes, underscoring the need for human oversight in the creative process. This emphasis on quality and control reflects his commitment to maintaining high standards in VFX, where AI serves as a supporting tool rather than an autonomous creator.

For artists interested in AI-driven VFX, Kahn emphasizes the importance of mastering foundational skills, such as composition, lighting, and perspective. He argues that these basics are crucial for using AI effectively, as they enable artists to direct AI tools with precision and intention. "Without a background in composition or lighting, you're going to hit a wall," he cautions, advising newcomers to build a strong technical foundation before fully embracing AI. Kahn sees this knowledge as essential for creating visually coherent work, particularly in a field where AI-generated outputs can lack consistency. In addition to technical skills, Kahn encourages artists to experiment with a variety of AI tools to find the right balance between automation and artistic control. He recommends Photoshop as a valuable resource, especially for refining AI-generated images and ensuring they meet project specifications. "Photoshop is still my best friend," he states, noting that it allows him to adjust AI outputs to achieve the specific visual effects needed for each shot. He has thus launched LoneLion Academy (https://www.lonelion.ca/) to train a new generation of filmmakers with his comprehensive masterclass on filmmaking with the technology.

Looking to the future, Kahn is excited about the potential for interactive media that adapts in real time to user input—a concept he believes will redefine storytelling. He envisions a world where films could evolve dynamically, offering viewers the ability to choose from "16,000 different endings" rather than the limited options available today. This concept aligns with a shift from static file formats to data-driven media, where video content can be customized on the go. "It's about creating a system where the viewer has control," he remarks, describing a vision where AI empowers audiences to engage with media in a participatory way. This transition to interactive, data-driven media represents what Kahn sees as the next frontier for AI in entertainment.

For Kahn, this future holds both opportunities and challenges, as it requires a new approach to content creation that blends technical skill with narrative flexibility. He is particularly interested in exploring how AI can facilitate this shift, providing tools that enable artists to create adaptable, responsive media experiences. His focus on interactivity reflects a broader trend in AI-driven video, where creators are moving beyond static content to embrace immersive, user-centered storytelling. This vision highlights Kahn's commitment to innovation, as he seeks to push the boundaries of what is possible in the evolving landscape of AI and media.

4.2.6 Guillaume Hurbault (France): Pioneering AI-Driven Cinematic Storytelling Beyond Traditional Filmmaking

Guillaume Hurbault, a pioneering AI filmmaker from France, has redefined his creative journey through the integration of generative AI in video production (Hutson, 2024f). Originally from a background in the video game industry, where he focused on sales, marketing, and online distribution, Hurbault did not follow a conventional path into filmmaking. His passion for storytelling, combined with a fascination for the growing potential demonstrated in generative tools, led him to explore new creative avenues. "I have no film background prior to AI—so I really came to film because of AI," Hurbault explains, capturing the unique entry point that generative technology provided. Initially driven by an interest in science fiction, he began experimenting with AI tools about two and a half years ago, discovering a medium that allowed him to produce visually compelling content without traditional cinematic training. His journey reflects the democratizing power of AI, making high-quality film production accessible to those outside the conventional filmmaking world.

His early work with AI involved playing with image generators and animation software, where he began to visualize new storytelling possibilities. He recalls the excitement of creating visuals that once would have required extensive budgets or technical expertise, describing the appeal of the tools as "creative potential beyond what I could have imagined." Although his initial results were experimental, the work of Hurbault quickly gained attention within the creative community. Winning an AI filmmaking contest with a short film last Halloween marked his first professional recognition, providing validation for his unconventional approach and fueling his commitment to further explore AI in cinematic storytelling.

As Hurbault progressed, he started using more sophisticated AI tools, which enabled greater control over the production process. "With tools like MidJourney, I could start to control camera angles, lighting, and the overall mood," he shares, reflecting on the rapid improvements in AI capabilities that transformed his work from random image generation to structured visual storytelling. A defining moment in his career came with the release of his film *One of Us* (https://www.youtube.com/watch?v=WeIVNsyUOJc) (2023) (Fig. 4.6), an emotionally resonant piece that diverged from typical AI-generated sci-fi or trailer-style content. Shot from the perspective of a newborn child in a middle-class American home experiencing his parents watching over him in the crib to learning to draw and experiencing the world, ending with the aftermath of a war. "It was a serious topic with a grim ending—not the usual epic or fun AI style," Hurbault recalls, noting how the project gained attention from industry professionals, including directors and VFX artists, who recognized the depth AI could bring to cinematic work. This breakthrough underscored his belief that AI could handle nuanced storytelling, positioning him as one of the first to build a career in AI-driven film.

For Hurbault, the role AI plays in film is not merely a technical aid but a powerful creative partner. While early projects often involved adapting stories to the limited,

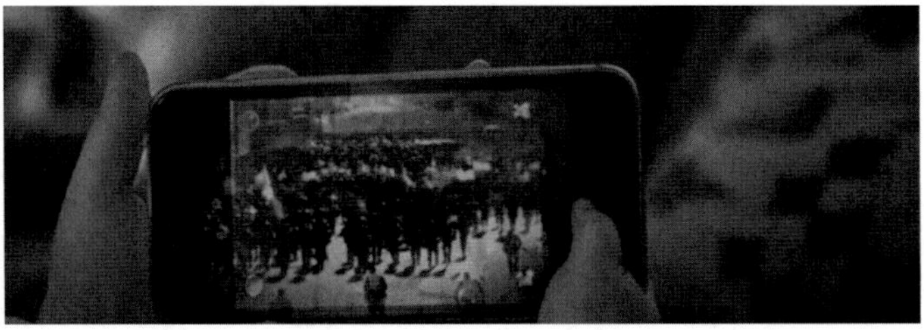

Figure 4.6 Guillaume Hurbault, *One of Us*, film still, 2023. Courtesy of Guillaume Hurbault.

random outputs the nascent models generated, advancements in tools have allowed him to reverse that process, creating stories with defined visual and narrative arcs. Describing this transition, he notes, "The tools now make me feel like I'm actually directing, not just compromising with the AI." This evolution has led Hurbault to shift his focus toward more complex projects, where AI serves as a means to realize his artistic vision rather than merely assisting with the technical workload.

Alongside Canadian co-founder Dale Williams, Hurbault established Solo Twin Studios (https://www.solotwinstudios.com/) to accommodate the growing demand for high-quality AI-generated films. While Hurbault brings a distinctively European, artistically ambitious perspective, Williams contributes a North American entertainment sensibility, making their collaboration particularly dynamic. "Dale has this strong entertainment background, while I bring a more artistic vision," Hurbault explains, describing how their complementary styles enhance the output of the studio. Their shared background in tech and video games also equips them with a practical understanding of digital media production, an asset in the fast-paced, adaptable world of AI filmmaking. Together, they have produced nineteen original series, along with music videos, advertisements, and other content that have reached millions across social platforms.

The establishment of Solo Twin Studios marked a new phase in Hurbault's career, transitioning from solo experimentation to a collaborative production model capable of handling larger projects with quicker turnarounds. This setup allows the studio to source top AI talent and meet the demands of diverse clients while maintaining the experimental spirit that characterizes Hurbault's work. "We're able to operate almost 24/7," Hurbault notes, with the team's international structure allowing constant progress on projects. The success of the studio demonstrates the scalability of AI-driven film production and reflects his vision for a future where small, adaptable teams create impactful visual stories.

Despite his excitement, Hurbault is mindful of the ethical implications AI brings to the creative industry. He acknowledges that AI allows him to work remotely, outside traditional media hubs like Hollywood or Paris, disrupting traditional power dynamics in film production. However, he also reflects on the potential impact on established industries and the risks of job displacement. "I'm not going to stop being creative just because it might hurt an industry," he states, adding that the challenges facing Hollywood are not solely due to AI. His perspective aligns with a broader view that AI is accelerating pre-existing shifts within the entertainment industry rather than introducing wholly new disruptions. Still, he remains sensitive to the need for responsible AI use, particularly as it gains traction in commercial applications.

This ethical stance reflects his desire to contribute meaningfully to the future of cinema. As he continues to develop his craft, he actively pursues traditional cinematic skills—such as storytelling, pacing, and editing—that AI cannot replace. In his view, the core elements that define great storytelling are timeless, transcending the specific tools or mediums used. "Whether you're painting in a cave or making a film with AI, a good story and pacing are essential," he says, underscoring his commitment to storytelling fundamentals. This dedication to craft, combined with a deep understanding of the capabilities, places Hurbault at the forefront of a new generation of filmmakers.

In envisioning the future, Hurbault is particularly excited about creating hybrid production crews that integrate AI and traditional filmmaking roles. He imagines small teams of five to ten people combining AI specialists, screenwriters, VFX artists, and other creatives to produce high-caliber, feature-length films. "The goal is to have people watch a film and say, 'This is amazing,' without qualifying it with, 'for AI,'" he explains. He sees this collaborative approach as the key to scaling production quality, allowing AI filmmakers to undertake ambitious projects previously restricted by budget or access to resources. The plan includes roles familiar to traditional cinema but adapted for AI workflows. For example, an AI casting director might be responsible for both generating characters and voicing them, while a screenwriter could handle not only dialogue but also the nuances of AI-driven character interaction.

This reimagining of crew roles emphasizes efficiency while preserving the artistry and detail that make storytelling compelling. As Solo Twin Studios grows, Hurbault is committed to pioneering this hybrid model, seeking to establish new industry standards that reflect the unique potential and challenges of AI in film. Through his work and vision for the future, Guillaume Hurbault exemplifies the transformative power of AI in storytelling, balancing technical innovation with a reverence for narrative craft. His approach points to a future where AI enables both the expansion of creative possibilities and the establishment of new industry models, ultimately enriching the cinematic experience for creators and audiences alike.

4.2.7 Simon Graff (Germany): Bridging AI, Spatial Computing, and Immersive Media

Moving further east to Germany, Simon Graff (https://simongraff.de/), an emerging tech professional from Hamburg, Germany, shares his journey through the realms of generative AI, spatial computing, and the Metaverse. Founder of the agency FOR REAL?! and chairman of nextReality.Hamburg e.V., Graff focuses on helping corporations and creative professionals understand and integrate new digital realities (Hutson, 2024g). His expertise ranges from consulting on the latest AI-driven tools to providing strategic guidance on the role of spatial computing in industries such as marketing, media, and education. His work reflects a commitment to bridging the gap between technological potential and real-world applications, a dynamic Graff describes as "getting hands-on with emerging tech to offer clear perspectives on what works, what doesn't, and where the hype needs to be reined in."

The professional journey of Graff began in 2013 when he explored virtual reality (VR) and spatial computing, using immersive environments to analyze user perceptions. Early on, he was struck by the transformative power of VR, especially in the way it generated presence—a concept that became the basis for his university thesis. From there, he transitioned into roles that allowed him to introduce VR and AI to a wider audience, educating clients on how emerging tech could reshape traditional workflows. "It's about getting companies to look past the buzz and see practical impacts," he explains, underscoring his focus on pragmatic, industry-specific solutions. His move into AI came later, sparked by his exposure to generative tools like ChatGPT and image generators, which opened new avenues for his projects.

Interestingly given his interest in emerging technologies, Graff began his journey with generative AI cautiously, initially focusing on AI for experimental and consultative purposes. His "moment of revelation" came with ChatGPT, which he encountered while on vacation. "It was mind-opening," he reflects, emphasizing how the conversational quality of ChatGPT surpassed prior chatbot technologies. From there, he delved deeper, incorporating AI into a variety of creative projects, from generating visual assets to developing a virtual fashion brand, Dianesis. This exploratory approach to generative tools has become a hallmark of his work, pushing the boundaries of what AI can do across domains, from art and fashion to immersive storytelling.

Today, Graff uses AI tools extensively, integrating platforms like MidJourney, Runway Gen-3, and Luma AI for tasks that range from prototyping to content creation. In his consulting work, he advocates for AI as a tool for efficiency, encouraging companies to view generative tech as an enhancer rather than a replacement. His clients, however, often approach AI cautiously, especially in sectors like beauty and fashion, where authenticity is paramount. Many brands remain skeptical about AI's ability to produce "real" content, expressing concerns over ethics and compliance with European data privacy regulations. Graff sees this hesitation as part of a broader European caution regarding new tech. "For many German brands, AI raises questions about compliance and the potential dilution of human authenticity," he notes, highlighting a cultural difference in tech adoption between Europe and other regions.

Despite these reservations, Graff believes AI's benefits for corporate workflows are clear. He recounts a project where a beauty brand tested AI-generated models for packaging shoots, aiming to save on costly logistics associated with live models. Although initial responses were mixed, a follow-up field study revealed that most viewers couldn't distinguish AI images from photos of real models. This finding, Graff believes, signals a turning point in AI's acceptance, particularly as companies recognize the cost-saving potential AI brings to repetitive and resource-intensive processes. His approach reflects a balance between innovation and practicality, as he navigates the ethical considerations while demonstrating AI's tangible value in professional settings.

In his role as chairman of nextReality. Hamburg, Graff is well-versed in the regulatory environment shaping AI and emerging tech in Europe. He explains that the EU's AI Act, which will be fully enforceable by 2027, has already begun to influence corporate AI adoption, with many organizations hesitant to fully integrate AI without clear regulatory guidance. “There’s a transition phase happening,” Graff remarks, emphasizing that while the AI Act isn’t yet fully binding, companies are preparing for compliance by 2027. This regulatory caution is reflected in the way European firms, particularly those in Germany, approach AI-driven innovation—often opting for conservative applications in anticipation of future legal constraints.

Graff contends that while regulatory caution is necessary, it shouldn’t impede creativity or experimentation. He notes that the AI Act primarily targets transparency in data practices for major tech companies like Meta and Google rather than individual creators or small businesses. “Many overthink it,” he observes, pointing out that AI’s potential to streamline and enhance creative work is too significant to ignore. He advocates for a balanced perspective, one that allows businesses to explore AI’s possibilities while maintaining ethical standards. This pragmatic approach has become integral to his work at FOR REAL?!, where he consults with clients on adopting AI in ways that align with regulatory expectations without stifling innovation.

Among his many projects, Graff’s virtual fashion label, Dianesis, serves as a showcase for the creative potential of generative AI. Designed as a conceptual experiment rather than a profit-driven venture, Dianesis demonstrates how AI can facilitate rapid, high-quality content creation. Graff describes the brand as “a demo tape for AI,” noting that it has garnered media attention despite being a side project. The brand’s visual identity was developed through ChatGPT and MidJourney, with AI providing both the creative vision and technical execution for virtual collections. Although still in the exploratory phase, Dianesis offers a glimpse into what AI-enabled fashion could look like in the future, where digital identities and metaverse-ready designs converge.

Graff’s ultimate vision for Dianesis extends beyond static 2D images. He envisions a future where the brand offers immersive, interactive experiences, allowing users to wear virtual designs in AR environments or within metaverse platforms. “It’s about creating a fashion label that feels tangible but lives entirely in the digital realm,” he explains. Graff’s work in virtual fashion highlights his forward-thinking approach, using AI to bring concepts to life that might otherwise remain abstract. His goal is to develop designs that can exist seamlessly across digital and real-world spaces, pushing the boundaries of what it means to “wear” fashion in a technologically mediated world. Venues such as AI Fashion Week (https://fashionweek.ai/) offer exposure for his work on the brand (Fig. 4.7).

Looking to the future, he speculates on the role of the Metaverse in storytelling, particularly as AI enables hyper-personalized content. He foresees a shift toward on-demand narratives, where viewers can request customized stories starring their favorite characters in unique scenarios. However, he remains skeptical about fully personalized cinema, questioning whether viewers truly desire unlimited control over content. “Sometimes, people just want to be entertained without making choices,” he reflects, suggesting that passive consumption will still have a place in a world of interactive media. This perspective underscores his nuanced understanding of the

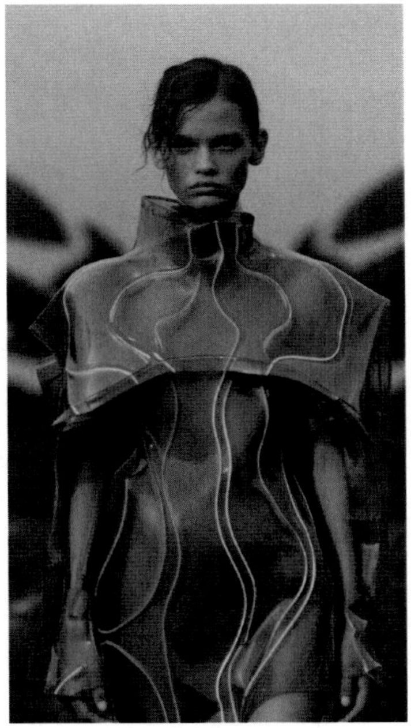

Figure 4.7 Simon Graff, *Runway Model*, AI Fashion Week, 2024. Courtsey of Simon Graff.

potential of AI, where technology expands possibilities without necessarily replacing established forms of engagement.

Through his work, Simon Graff exemplifies a balanced approach to AI and emerging tech, combining visionary concepts with grounded, practical applications. His insights into the evolving landscape of digital media offer a thoughtful perspective on the opportunities and challenges AI presents, from regulatory concerns to creative possibilities in the Metaverse. His journey demonstrates the transformative power of technology when approached with both ambition and caution, shaping a future where AI and human creativity coalesce in dynamic, innovative ways.

4.2.8 Yonatan Dor (Germany): Advancing Visual Storytelling with AI in Art and Advertising

Moving from Hamburg to Berlin, we explore the work and journey of Yonatan Dor, a Berlin-based AI video artist and creative director of The Dor Brothers (https://www.thedorbrothers.com/) who has pushed the boundaries of AI-generated visuals in the global entertainment and advertising industries (Hutson, 2024h). Dor began his career with traditional music and video production, capturing live-action footage with practical equipment and working closely with musical artists to produce compelling visuals. Over time, however, Dor found himself constrained by the high costs and logistical demands of conventional filmmaking. "I wanted to create these grand visuals, like gods battling on Olympus," he recalls, reflecting on how his expansive creative vision often collided with the limitations of traditional production. As he searched for new ways to realize his ideas without the prohibitive costs,

Dor discovered generative AI in 2022, diving into platforms like Stable Diffusion and MidJourney, becoming one of the earliest adopters of these groundbreaking technologies. This shift transformed his workflow and opened up a world of creative possibilities that had previously been out of reach.

Given his early professional work in the world of music videos, he worked closely with artists to produce visually compelling narratives that complemented their sound. This experience helped him develop an eye for visual storytelling and an understanding of how to translate abstract themes into concrete, cinematic imagery. In his early days, Dor experimented with various tools, including 3D animation software like Unreal Engine (https://www.unrealengine.com/en-US), which allowed him to construct elaborate virtual scenes. However, he quickly encountered the labor-intensive nature of such projects. "I realized that creating a 10-second scene could take a month," he notes, describing the immense time investment required to achieve even brief sequences. The slow pace and technical demands of traditional animation and 3D software made him reconsider his approach, particularly as his ideas grew more ambitious.

This quest for efficiency and creative freedom eventually led Dor to explore generative AI. In July 2022, he began experimenting with Stable Diffusion and August with MidJourney, captivated by the potential to generate complex visuals directly from text prompts (Fig. 4.8). Despite the early limitations of these models, which often produced unpredictable and abstract results, Dor was drawn to the possibilities they offered. "The fact that AI could take text and turn it into visuals was astonishing," he reflects, highlighting the novelty and accessibility of these tools compared to traditional methods. Generative platforms allowed him to visualize grand concepts that would otherwise be cost-prohibitive, marking the beginning of his shift from traditional media to AI-enhanced production.

Through continuous experimentation and refinement, Dor developed a unique aesthetic that combined the strengths of various AI tools to create cohesive, high-quality visual narratives. He describes his creative process as an integration of multiple tools, each chosen for its ability to handle specific aspects of a scene. Some platforms excel at generating lifelike movement but may lack depth, while others produce static images with a high level of detail. By layering these outputs and using post-processing techniques, Dor has crafted a distinct style that feels both innovative and immersive. "It's about using AI as a creative partner, not a substitute for human vision," he explains, underscoring his belief in such potential to enhance rather than replace artistic intent.

This methodology has allowed Dor to produce over 300 music videos, including high-profile collaborations with internationally recognized artists such as French Montana, SiM (https://youtu.be/ow1KL6u_08c), and INSOMNIUM. These projects have garnered millions of views, solidifying his reputation as a pioneer in AI-generated visuals. For Dor, each project is a new opportunity to push the boundaries of what can be achieved, merging the technical precision of these tools with his own narrative instincts to create visuals that resonate with audiences on an emotional level. His work has helped redefine what is possible in AI-enhanced video production, positioning him at the forefront of this rapidly evolving field.

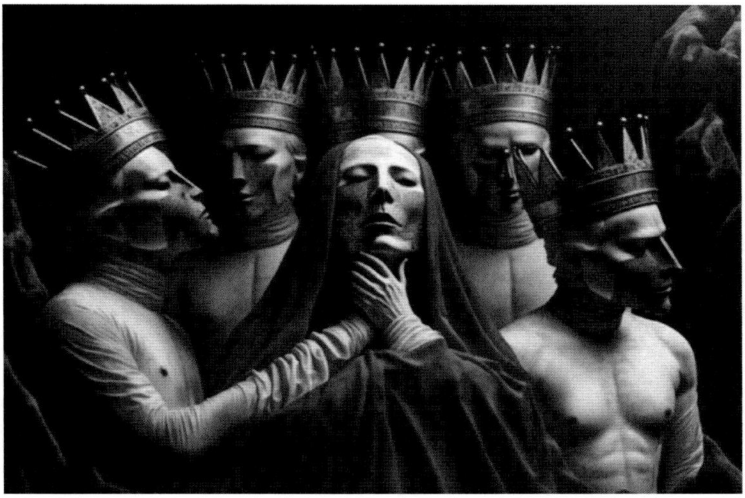

Figure 4.8 The Dor Brothers, *Voluntary Submission*, Midjourney, 2022. Courtesy of Yonatan Dor

As Dor's influence in the AI video industry grew, so did the ethical questions surrounding the use of AI-generated content. He approaches these challenges with a commitment to transparency, always openly labeling his work as AI-generated to avoid misleading his audience. "I never hid the fact that I'm using AI," he asserts, emphasizing his dedication to ethical standards in an industry where AI's role is often ambiguous. In *The Hustle* (https://www.youtube.com/watch?v=aIxqms8KSkA&t=1s) (2024) (Fig. 1.2), Dor uses political satire to underscore the importance of transparency in AI-generated media, labeling the video as AI-created to highlight the ethical considerations around deepfakes and generated content. The film portrays a series of high-profile political figures, including Donald Trump, Kamala Harris, Joe Biden, and Barack Obama, engaging in humorous yet surreal scenarios. Through this portrayal, Dor emphasizes the potential for AI to blur reality, demonstrating how the technology can create convincing, yet entirely fabricated, footage of public figures. By openly disclosing the use of AI in *The Hustle*, Dor invites audiences to reflect on the ethical implications of AI in media, especially as it relates to political imagery. His satirical approach acts as a warning about the need for careful discussions on regulation and transparency, asking society to consider the consequences of AI-generated content without clear labeling. The work thus serves as both entertainment and a commentary on the broader risks associated with misrepresentation, reinforcing his advocacy for responsible AI practices in an era where technology's role in shaping public perception is rapidly expanding.

As AI gains traction in commercial applications, particularly in advertising and branded content, Dor has become an advocate for responsible AI use, stressing the importance of clear disclosures and ethical practices. "We're not here to take people's jobs," he says, "but to create a new market that integrates AI with human creativity." This balanced approach reflects his view that AI and traditional artistry can coexist, each enhancing the other in a collaborative ecosystem. The ethical stance extends to his hiring practices, as he continues to work with traditional illustrators, screenwriters, and VFX artists, integrating their expertise into his AI-driven workflow. By blending human talent with AI capabilities, he aims to create a model where technology supports rather than replaces human input. His collaborations with various artists demonstrate his commitment to fostering a creative environment that values both technological innovation and artistic integrity, positioning his work as a testament to the potential for AI to enhance rather than undermine traditional creative fields.

Beyond video production, Dor has expanded his influence into the realm of interactive art through projects like Event Station, a physical photobooth that offers AI-generated, personalized artwork. This venture merges technology with experiential art, providing users with unique, custom images created in real-time by AI. Event Station reflects the interest in blending physical and digital experiences, allowing people to engage with AI as both a creative tool and an interactive medium. The photobooth has proven popular at events and exhibitions, offering an accessible introduction to AI-generated art for the general public. "It's about bringing AI to the people," Dor explains, viewing Event Station as a bridge between everyday audiences and cutting-edge technology.

This project is part of a broader vision to make AI art more accessible and engaging, a mission that aligns with Dor's participation in the art team at Pika Labs and his presentations at industry conferences, including sessions for Microsoft. Through these collaborations and public appearances, Dor has positioned himself as a thought leader in AI art, using his platform to educate others on both the creative potential and ethical considerations of generative technology. His work at conferences and partnerships with high-profile clients demonstrate his commitment to elevating the dialogue around AI in creative industries, highlighting its possibilities while addressing its challenges.

The expertise of the artist extends into high-caliber commercial projects for prominent brands, with advertisements displayed in major locations such as Milan's city streets and during internationally watched football matches. These projects showcase his ability to apply AI-driven visuals to large-scale, mainstream campaigns, demonstrating the commercial viability of generative technology. Working with high-profile clients like Tesla (https://youtu.be/oBzYYAgq8dE), Yamaha, and more has allowed Dor to blend his artistic vision with commercial appeal, creating visuals that captivate diverse audiences. His success in the commercial sector underscores the adaptability of AI-generated art, proving that it can stand alongside traditional media in terms of quality and impact. "AI allows us to bring complex visual ideas to life quickly and effectively, something that's invaluable in advertising," he states, reflecting on the efficiency AI brings to high-demand environments.

Looking forward, Dor envisions a future where AI evolves to directly translate human thoughts and ideas into visual form, a concept he refers to as "thought-to-canvas." He believes that as AI continues to develop, the technology will become indistinguishable from traditional media, allowing artists to focus more on conceptual development and less on technical execution. "In the end, what matters is your idea," he emphasizes, suggesting that AI's true value lies in its ability to amplify creativity without overshadowing the creator's voice. This vision aligns with recent advances in human-computer interaction, where AI could eventually enable real-time visualization of human thoughts, bridging the gap between imagination and tangible output.

Dor is cautiously optimistic about these developments, recognizing both the immense creative potential and the ethical complexities they entail. As AI becomes more integrated into the artistic process, he advocates for a balanced approach that respects the role of human oversight. His commitment to responsible AI use and his openness to collaboration with traditional artists highlight his dedication to a future where technology and creativity coexist harmoniously. Through his innovative approach and his willingness to engage with the ethical questions surrounding AI, Yonatan Dor exemplifies the transformative power of AI in visual storytelling, leading the way toward a future where AI serves as a tool for expanding—not replacing—the possibilities of artistic expression.

4.2.9 Christian Georgiev-Fries (Austria): Integrating Traditional Art Principles into AI

Christian Georgiev-Fries (https://freezephotos.wordpress.com/), an AI video artist based in Vienna, Austria, has built an extensive career that blends traditional and digital art practices (Hutson, 2024i). His journey began in analog photography, evolving through the dawn of digital media around 2004 and 2005, and deeply influenced by the rise of mobile and social media technologies. With a foundation in fashion photography, commercial studio work, and even gallery experience with the Young British Artists (YBA), Georgiev-Fries developed a unique aesthetic grounded in traditional art principles. "Growing up with these shifting mediums made me more adaptable," he recalls, noting that this adaptability has been essential in navigating the current technological landscape of generative AI. His early exposure to digital tools like Photoshop, starting in 1999, laid the groundwork for his digital ventures, which have spanned from commercial advertising to more experimental projects in the art world.

The transition of Georgiev-Fries into the digital sphere was facilitated by his engagement with a startup tech company, where he initially worked as a translator for English reviews and later moved into digital marketing. This experience marked his first significant foray into the digital world, allowing him to synthesize his background in studio arts with emerging technologies. "In 2016, I started seeing how digital advertising could leverage principles from studio art," he explains, describing how his foundation in painting, sculpture, and installation helped inform his approach to digital media. After a stint in Boston, where he continued developing skills in video production and marketing, Georgiev-Fries returned to Vienna and became fully immersed in generative AI around 2022. With the release of DALLE and other generative tools, he saw parallels between AI and traditional tools, viewing AI as a "camera that lets me focus on concepts rather than the technical aspects."

Since adopting AI tools, Georgiev-Fries has leaned heavily into the narrative-driven potential of generative AI. He uses a "Gen stack" comprising

MidJourney, HeyGen, ChatGPT, and Claude, to create highly personalized, story-rich visuals for commercial clients (https://youtu.be/xdc7TCtNQrw). This approach allows him to produce compelling B-roll and voiceovers without relying on extensive traditional setups. "My goal is always to have a clear narrative, a story that drives the visuals," he notes, emphasizing that AI has freed him to focus on storytelling rather than technical execution. By employing AI in a way that mirrors the artistic decisions he would make in traditional photography or filmmaking, Georgiev-Fries crafts a workflow that seamlessly integrates conceptual intent with technological efficiency. This balance of creative vision and practical execution is particularly valuable in commercial projects, where he sees AI as a tool to streamline processes without compromising quality.

Georgiev-Fries attributes much of his success with AI tools to his deep understanding of artistic fundamentals like form, composition, and narrative structure. "It's about knowing the difference between subjective and objective perspectives," he explains, adding that his fine art background allows him to see AI-generated visuals through a refined, critical lens. He draws from his early training in gallery spaces and studio arts to approach each project with a rigorous sense of form and color composition, which he applies even in AI-driven contexts. His work, therefore, straddles the line between fine art and commercial art, appealing to audiences familiar with both domains. He believes that this nuanced approach is essential for creating work that resonates emotionally while meeting the technical demands of the digital landscape. "Every tool has its place," he says, likening AI tools to the way painters understand their brushes and pigments.

A key concern for Georgiev-Fries is the accessibility of generative AI tools, which he sees as a barrier to entry for many aspiring artists. He notes that while some tools offer free versions, there is often a substantial difference in quality between these and paid versions. "Not everyone can afford the $100-a-month Gen stack," he points out, arguing that access to high-quality tools is an equity issue in the art world. This concern extends to ethical questions about copyright and intellectual property, which he navigates by using licensed assets and creating original prompts. "If I'm drawing from stock, it's always licensed," he asserts, mindful of the potential legal complexities surrounding AI-generated content. His conscientious approach reflects a commitment to both artistic integrity and legal accountability, positioning him as a responsible innovator in the field.

Looking ahead, Georgiev-Fries is optimistic about the potential to enhance both commercial and fine art. He envisions AI becoming more integrated into creative workflows, enabling artists and marketers alike to produce more refined, narrative-driven content. His commercial projects, which include short-form videos optimized for platforms like YouTube, capitalize on the ability to produce engaging, high-impact visuals quickly and efficiently. For him, the future of AI in art lies in its capacity to complement human creativity, allowing artists to execute their visions with greater speed and precision. In this vision, AI is not a replacement for traditional art skills but rather a tool that augments and enhances them, empowering creators to push the boundaries of visual storytelling in innovative ways.

4.2.10 Luka Tisler (Slovenia): Pioneering Hybrid VFX and AI Methods

Luka Tisler (https://lukatisler.com/), a VFX artist and AI video creator based in Ljubljana, Slovenia, has crafted a unique journey through the intersection of traditional visual effects, 3D animation, and generative AI (Hutson, 2024j). With nearly two decades of experience in video production, VFX, and compositing, Tisler embodies the transition from analog to digital to AI-driven creative workflows. In a region where technological adoption in creative industries can be slow, Tisler has become a pioneer, navigating what he describes as "a landscape that's slowly getting into mainstream." Reflecting on his role, he notes, "I kind of feel like a lonely alien here, but I'm working hard to bring Slovenia forward in this field." His career trajectory captures the shifts within the media industry, from traditional compositing to cutting-edge AI, making him a valuable voice in AI-driven visual storytelling.

The early career of Tisler involved working behind the camera, producing, editing, and compositing video content. His experience spans both traditional techniques and VFX, utilizing tools such as Adobe After Effects, Fusion, Nuke, and Cinema 4D. However, the evolution of his creative process reached a pivotal moment with his introduction to AI-based tools about two years ago. Initially working with video-to-video applications in Runway Gen-1, Tisler saw the transformative potential AI brought to visual production. "It was amazing but not quite there yet," he recalls, acknowledging how each generation of AI tools offered new possibilities for creative expression. His "aha moment" came with Runway Gen-2 and later Gen-3, which he describes as "game-changing," particularly for its capabilities with motion brush and video generation, marking a turning point in his artistic approach.

As AI tools advanced, Tisler began experimenting with a hybrid model, combining traditional VFX techniques with AI-generated imagery. This approach allowed him to layer AI outputs over hand-crafted VFX, producing work that merges the polish of traditional effects with the fluidity and speed of AI. For Tisler, generative AI enables a new level of efficiency, especially in creating background scenes and establishing shots that might otherwise require extensive resources. He explains, "AI has democratized storytelling; now, you only need your imagination and a phone to bring grand ideas to life." This accessibility is essential for Tisler, who views AI as a tool that allows more people to tell stories without the constraints of a large production budget.

One of his core beliefs is the importance of curiosity and a willingness to learn across disciplines. Rather than specializing narrowly, he sees value in combining skills from various areas—visual effects, 3D animation, and AI video—to enhance his creative toolkit. "I'm a Jack of all trades, and finally, that's starting to pay off," he reflects. His approach involves using multiple AI tools in tandem, pulling images from MidJourney, generating videos in Runway, and compositing them in Adobe After Effects or Nuke to achieve the desired cinematic quality as seen in his 2024 AI showreel (https://www.youtube.com/watch?v=jbDuHZWn-50). This method provides him with flexibility and control, allowing him to craft visuals that resonate with audiences while meeting the technical demands of modern VFX.

The process of the artist typically involves generating a sequence of images or video snippets using various AI tools, iterating on them, and then assembling the final piece in post-production. This hybrid approach, which he calls "playing with tools to see what happens," has enabled him to produce a diverse range of work, from short films to commercial projects. His workflow embodies the spirit of experimentation, with AI offering him new avenues to explore narrative depth and visual complexity. "Experimentation is key," he says, emphasizing that the ability to integrate AI into one's creative process is as much about attitude as it is about technical skill.

Working with clients unfamiliar with AI capabilities presents its challenges, and Tisler spends a significant portion of his time educating them on the potential and limitations of generative technology. "Most clients have no idea what these tools can do," he explains, noting that managing expectations is essential. His approach involves showing clients examples of AI-generated work and discussing what is feasible within their project's scope. He emphasizes the importance of a "fluid script," advising clients to think creatively and be open to adjusting their vision to align with AI's capabilities. "Sometimes you can't get the exact shot you imagine, but AI allows you to find alternative ways to convey the same message," he says, highlighting the need for flexibility in AI-driven production.

Tisler also encourages clients to adopt an "AI mindset," understanding that generative tools operate under different constraints than traditional cameras and crews. He frequently collaborates with clients on iterative projects, explaining the limitations of current AI tools while offering creative solutions. For Tisler, this collaborative, back-and-forth approach helps bridge the gap between traditional expectations and AI's unique creative potential. His educational role within these partnerships not only demystifies AI but also fosters a deeper understanding of how AI can be used to create impactful, narrative-driven content.

Looking to the future, Tisler hopes that more professionals in traditional VFX and 3D animation will embrace AI and incorporate it into their workflows. "There are so many talented people in the old paradigm who could do incredible things if they'd just give AI a try," he observes. He understands, however, the resistance that some established artists feel toward adopting new tools, especially after years of mastering traditional techniques. Yet Tisler believes that AI's potential to expand creative possibilities is too significant to ignore, and he hopes to see a more inclusive community where both AI-based and traditional skills are valued. His vision for the future involves a collaborative, open-minded approach to VFX, where artists can leverage AI without compromising the artistic integrity honed through years of experience.

The commitment of Tisler to advancing the use of AI in visual storytelling reflects a broader aspiration to make high-quality content creation accessible to more people, regardless of their technical background or financial resources. Through his work and his advocacy for the potential of the tools, Luka Tisler is helping to build a bridge between traditional VFX practices and the transformative capabilities of AI, guiding the industry toward a future where creativity, technology, and inclusivity come together in powerful ways.

4.2.11 Aashay Singh (India): Fusing Spirituality and Technology in AI Filmmaking

Aashay Singh (https://www.youtube.com/@aashay), an AI video artist and filmmaker based in southern India, has carved a unique creative path blending technology, storytelling, and a deeply spiritual outlook (Hutson, 2024k). Known for his innovative use of generative AI in video, his work reflects a philosophical depth, rooted in his personal journey from traditional filmmaking to the forefront of AI-driven art. His approach diverges from conventional artistic norms, as Singh considers his work an expression of personal joy and a reflection of his spiritual beliefs rather than a rigid statement of intent. "Our bodies are prisons for our souls," he quotes, conveying a worldview that permeates his art, where life and art are seen as transient and interconnected experiences. This philosophy underpins his exploration of AI as a tool not just for creation, but for realizing an expressive, boundary-breaking approach to visual storytelling.

The journey of Singh began with a profound ambition to "save the world" upon completing high school, an aspiration that led him to study at a United Nations program and later pursue computer animation and game design in England. However, his exposure to global settings revealed to him the often superficial responses to real-world issues, prompting him to refocus his goals. This experience influenced Singh's shift from external activism to internal, personal exploration, ultimately steering him towards the expressive potential of filmmaking. His early studies and projects in animation and filmmaking helped Singh hone his technical skills, yet he felt constrained by conventional artistic pathways. "Growing up in India, you don't even consider saying you want to be a filmmaker," he reflects, noting the societal pressures to pursue more traditional career paths. His time in England, however, offered a liberating environment, encouraging him to switch to filmmaking and explore storytelling through visual media.

Returning to India marked a turning point as Singh embarked on a spiritual journey, drawing from Hindu and Buddhist philosophies and integrating meditation into his life. This shift not only informed his worldview but also his artistic approach, as he sought to create art that resonates with the transient and introspective nature of human experience. Over the years, Singh returned to filmmaking, using his technical skills in 3D animation and visual effects to experiment with new forms of visual storytelling, a journey accelerated with the advent of AI technology in recent years.

The early cinematic works of Singh, such as *Supramental: A Kino Kabaret Short Film* (https://www.youtube.com/watch?v=iRyQosUaxR8) (2017) (Fig. 4.9), reflect his roots in traditional filmmaking and a fascination with existential themes. Created within a 50-hour timeframe for the Kino Kabaret challenge, the short film showcases Singh's ability to evoke profound questions through minimalist production. Shot on a GH2 camera with the Sanity hack and a 20 mm pancake lens, the film's stripped-down aesthetic complements its intense narrative, featuring only two actors in a confined, almost claustrophobic space. The storyline unfolds with a woman confronting a yogi, seeking the meaning of life. In a shocking turn, the yogi slits her throat with a fan, delivering a darkly ironic line, "Don't worry. Be happy... or not. You can worry as long as you want, but you're still dead." This juxtaposition

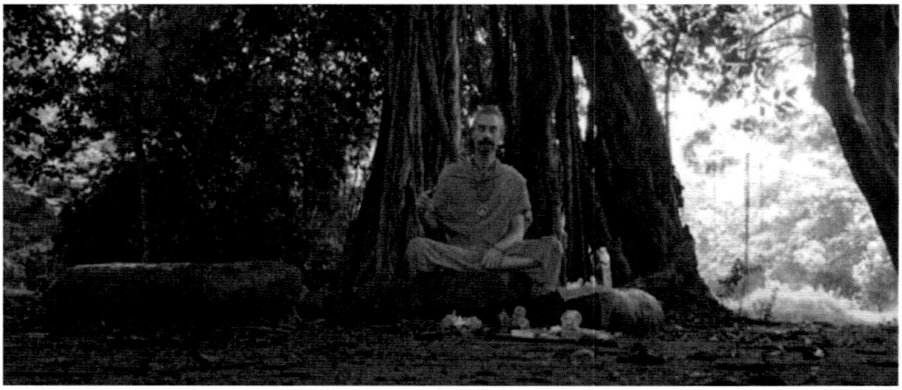

Figure 4.9 Aashay Singh, *Supramental. A Kino Kabaret short film*, film still, 2017. Courtesy of Aashay Singh

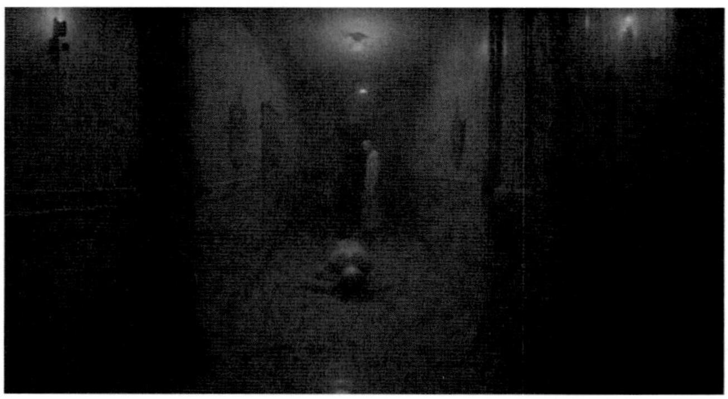

Figure 4.10 Aashay Singh, *Hotel Beksinski*, film still, 2024. Courtesy of Aashay Singh.

of humor and violence underscores Singh's bold storytelling style, pushing viewers to confront the uncertainties of life with an unflinching lens. These early works reveal an inclination toward philosophical exploration, setting the stage for his later, AI-enhanced narratives that continue to challenge conventional perspectives on existence and consciousness.

AI entered Singh's creative toolkit as a natural extension of his evolving artistic vision. Initially experimenting with free tools like ChatGPT and the servers of Hugging Face, Singh embraced the accessibility of these platforms, allowing him to create without the financial burden of traditional filmmaking setups. His early work, produced using minimal resources, attracted attention on platforms like Instagram, marking the beginning of his journey into AI video art. With the release of more advanced tools such as Runway Gen-2 and Pika, Singh began exploring the potential for more complex projects, layering AI-generated imagery with traditional editing techniques. This hybrid approach enabled him to create his first AI-driven film, *Hotel Beksinski* (https://www.youtube.com/watch?v=o3R2SPkb30U&t=38s) (2024) (Fig. 4.10) which was featured at the Amsterdam Film Festival—a significant milestone that validated his experimental approach. The haunting monochromatic montage of slow, panning sequences inside of a flooded hotel includes a foreboding soundtrack. As the short film unfolds, various humanoid forms emerge from the walls, writhe on the floor or sit on bathtubs as the ominous atmosphere is further supported with the ghost-like transitions common in surreal AI video.

The process of the artist involves a blend of AI-generated images and traditional post-production techniques, often using MidJourney, Runway, and Photoshop's generative fill to expand visuals into cinematic compositions. He describes AI as "a freeing tool," enabling him to focus on storytelling and aesthetics without being bogged down by technical limitations. "The tool presents what it can, and I work with that," he explains, emphasizing a collaborative, fluid relationship with AI. This adaptive approach has allowed Singh to create visually rich films that explore themes of isolation, confinement, and spirituality, often inspired by his favorite directors, such as Andrei Tarkovsky (1932–1986) and David Lynch (1946–).

While AI has expanded Singh's creative horizons, it has also introduced unique challenges, particularly in balancing client expectations with the limitations of generative tools. In his commercial work for agencies worldwide, Singh often finds himself educating clients on capabilities, managing expectations by explaining the strengths and limitations of the new technology. This balance between creativity and practical application requires Singh to adapt his workflow continually, using a mix of AI and traditional VFX techniques. For instance, he describes generating assets in AI and compositing them in After Effects, allowing him to produce consistent visuals that meet client specifications while leveraging the efficiencies of these platforms. "You have to guide the client on what's possible," Singh explains, noting that AI's evolving nature requires a flexible, communicative approach with clients unfamiliar with generative technology.

Singh has also encountered skepticism from traditional artists and filmmakers, who question the place of AI in creative fields. However, he views these reservations as an extension of rigid artistic paradigms, likening the resistance to religious dogma. "These concepts of 'good' or 'bad' in art are deeply rooted in a binary view of creativity," he observes, advocating for a more open-minded perspective. Singh believes that AI democratizes artistic expression, inviting both seasoned and emerging creators to explore new visual languages. His engagement with the supportive AI community, which he describes as "the most encouraging," has further solidified his belief in AI's potential as a tool for diverse storytelling.

His artistic philosophy diverges from traditional approaches that prioritize rigid artistic statements. For him, art is an expression of curiosity and exploration, where AI serves as a collaborative medium rather than a tool to enforce specific ideas. He frequently begins projects without a predefined concept, drawing inspiration from poetry, music, or spontaneous AI-generated imagery. "I start with something, see where it goes, and follow it," he says, underscoring a flexible approach that contrasts with the precision often associated with film production. This method aligns with his spiritual belief that creativity should reflect a fluid relationship with the world, where art emerges organically rather than through meticulous control.

In his work, Singh is drawn to creating "ethereal" or "preternatural" visuals, using AI's inherent unpredictability to evoke a sense of otherworldliness. These abstract elements are not accidental but are embraced as part of AI's unique aesthetic, resonating with Singh's interest in meditative storytelling. Inspired by Tarkovsky's long takes and Jodorowsky's surrealism, he aims to create immersive, contemplative films that transcend conventional narrative structures. Singh's work reflects a commitment to exploring the intangible aspects of human experience, often through stories of isolation, introspection, and transcendence.

For those interested in AI-enhanced filmmaking, Singh emphasizes the importance of a strong foundation in traditional storytelling and cinematic language. He advises aspiring creators to learn narrative structure, cinematography, and editing, skills that he believes are essential for effectively guiding AI's creative output. "AI is just another tool, but storytelling is universal," he remarks, suggesting that while AI simplifies certain technical processes, a thorough understanding of visual language remains indispensable. Singh encourages artists to experiment freely with AI, viewing it as a means to amplify rather than replace their creative vision. "Every innovation, from Blender to Unreal, felt the same—just tools that free you to focus on the story," he reflects, advising creators to embrace AI as a complementary tool within a broader, well-rounded skill set.

In the perspective of Singh, the true value of AI lies in its ability to democratize creativity, allowing artists to realize projects that might otherwise be financially or technically prohibitive. His experience illustrates how AI can bridge the gap between vision and execution, enabling creators worldwide to tell stories previously limited by resources. As Singh continues to push the boundaries of AI video art, his journey serves as a testament to the power of technology to unlock new dimensions of artistic expression, grounded in both personal exploration and professional innovation.

4.2.12 Peter Zarko-Flynn (Australia): Merging Traditional Storytelling with AI for Emotional and Historical Narratives

Peter Zarko-Flynn (https://www.youtube.com/channel/UCbHsXCGJjXkqSARjBsfy6jg), widely recognized under his professional alias PZF, is a British filmmaker and photographer now based in Australia (Hutson, 2024l). With a background steeped in film studies, theoretical cinema, and a career that began at the BBC, Zarko-Flynn brings a profound understanding of narrative and visual storytelling to his AI-driven work. He sees AI as a tool that amplifies storytelling potential, enabling him to merge his passions for filmmaking, photography, and technology. In his words, "All that we see, or seem, is but a dream within a dream," an Edgar Allan Poe quote he believes captures the essence of his AI explorations in cinematic forms. His work draws on influences from European post-war cinema and emphasizes emotional storytelling, blending old-world artistic principles with modern AI capabilities.

The journey into AI filmmaking began as an experiment but quickly evolved into a passionate endeavor for PZF. Partnering with Leonardo AI and RunwayML, he uses generative tools to create dramatic short films and music videos. Initially, he started with still images but soon embraced AI video generation when Runway introduced its motion feature. This new medium allowed Zarko-Flynn to leverage his cinematic knowledge and experience, combining his love for "*mise-en-scène* and lighting" with the efficiency of generative tools in creating immersive, narrative-driven films. AI tools enable him to explore concepts such as memory, emotion, and surrealism within a short timeframe, pushing him to "experiment with visuals and narratives that might otherwise be cost-prohibitive in traditional filmmaking."

One of his most notable projects is *Journey to Chernobyl* (https://www.youtube.com/watch?v=OMQo-n62qck) (2024) (Fig. 4.11), a piece he describes as transformative in his artistic journey. Leveraging AI to animate his photographs from a previous visit to Chernobyl, he recreated life in the exclusion zone before the nuclear disaster and uses haunting, abandoned imagery of the snow-covered locations that now exist as a reminder of the event. This work not only demonstrated the power of new tools to bring still images to life but also highlighted its educational potential. The piece resonated with audiences for its emotional depth and historical significance, marking a pivotal moment in the creator's career and validating his belief that AI can be a medium for impactful storytelling.

For Zarko-Flynn, navigating the evolving landscape of AI and intellectual property is a nuanced process. Early in his AI journey, he used familiar narratives, and public domain works to anchor his experiments, such as the reinterpretation of Poe's *A Dream Within a Dream* (https://www.youtube.com/watch?v=_YBggW0ZMO8) (2024) (Fig. 4.12). The sequence of ghostly and haunting images of young men and women juxtaposed with sinister old figures holding babies is set to slow pop music. This approach gave him the creative freedom to explore AI without overstepping boundaries, though he acknowledges the complexities around copyright in generative art. In his view, "AI can bring personal projects to life in a way that would have been impossible with traditional resources." By focusing on original concepts and drawing from his own photography, Zarko-Flynn sidesteps many of the ethical challenges around copyright, allowing his art to remain both innovative and respectful of artistic integrity.

The filmmaker also speaks to the broader issue of authenticity in AI-generated art. He recognizes that the AI-generated visuals may not always be perfect but appreciates the iterative nature of AI as part of the creative process. "People think we're trying to make Hollywood," he notes, "but it's more about experimenting and seeing how we can push this technology." For Zarko-Flynn, AI-driven storytelling isn't a quest for flawless output but an exploration of AI's capacity to evoke emotion, challenge norms, and expand cinematic possibilities.

With a background in photography and journalism deeply influencing his approach, he notes that his experience as a photographer gives him a unique perspective on composition, light, and framing, essential elements that he incorporates into AI-generated scenes. "Good photography informs good filmmaking," he says, explaining how his understanding of visual storytelling aids in guiding AI outputs toward cohesive narratives. This influence is particularly evident in pieces like *Sonata* (https://www.youtube.com/watch?v=TTWU47qN2o8&t=67s) (Fig. 4.13), a film inspired by his personal experience with loss, which employs music and symbolic imagery to communicate complex emotions without dialogue. Set to Bethoven's *Moonlight Sonata*, the narrative unfolds as a mystery of a nurse and a patient in hospice that seems to tell a tale of infidelity but reveals one of the remembered loss

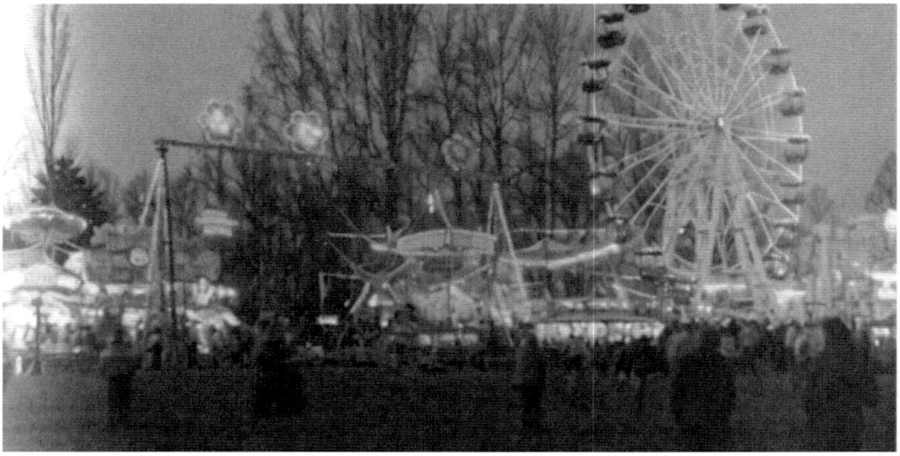

Figure 4.11 Peter Zarko-Flynn (PZF), *Journey to Chernobyl*, film still, 2024. Courtesy of PZF.

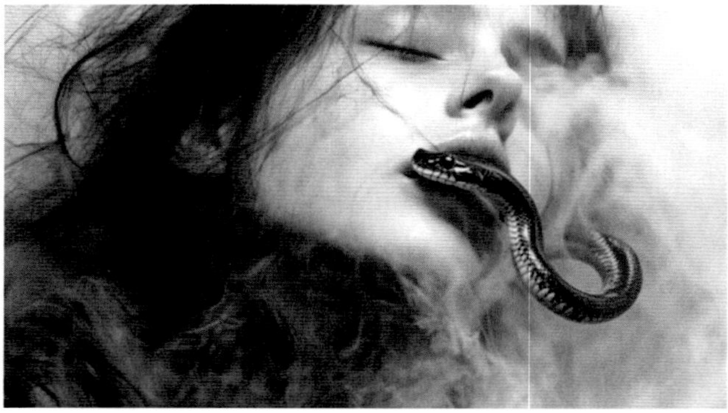

Figure 4.12 Peter Zarko-Flynn (PZF), *A Dream within a Dream*, film still, 2024. Courtesy of PZF.

by the end. The careful selection of visual elements by the director, and the layering of them with evocative soundscapes creates narratives that resonate emotionally, emphasizing the potential role as a complement rather than a replacement for traditional techniques.

His journalistic background also impacts his narrative choices, favoring stories with clear structures and relatable themes. Rather than adopting the more abstract or avant-garde approaches seen in much AI art, Zarko-Flynn maintains a traditional, linear narrative style. He believes audiences connect best with stories that unfold in a familiar structure, drawing parallels to storytelling practices that span human history. "Shamans, poets, and filmmakers alike have always known the power of a well-told story," he notes, adding that while AI offers new tools, the core of storytelling remains universal.

Looking ahead, Zarko-Flynn envisions a cinematic landscape where AI complements rather than replaces human creativity. He is hopeful that AI will democratize the industry, allowing independent creators to tell stories that might

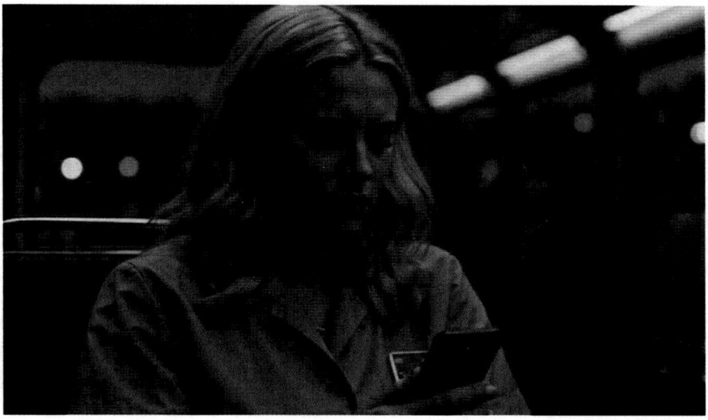

Figure 4.13 Peter Zarko-Flynn (PZF), *Sonata*, film still, 2024. Courtesy of PZF.

otherwise go unheard. However, he also recognizes the potential for oversaturation, expressing concerns about the "noise" that mass accessibility could generate. Reflecting on his experiences with media deregulation at the BBC, he sees parallels between the explosion of content on platforms like YouTube and the risk of diminishing quality in AI-driven films. "If everyone's a creator, who's left to watch?" he asks, underscoring the importance of maintaining quality and authenticity amidst technological advances.

This optimism about the educational potential of AI also fuels his vision for the future. He believes AI can be instrumental in preserving history, providing immersive experiences that make complex or forgotten stories accessible to new audiences. His work, especially in historical narratives, highlights AI's role in connecting viewers with the past in engaging and meaningful ways. Through his films, he hopes to inspire others to see AI as a medium for storytelling, one that enables personal expression while respecting the essence of human connection and experience.

4.3 Reflections on AI in Cinematic and Visual Storytelling

The series of interviews with AI-driven filmmakers and artists underscores the evolving impact of AI on contemporary visual storytelling, revealing recurring themes that resonate with the historical progression of technology in creative production (Table 4.1). As previously examined, the field of AI-generated art and cinema has transitioned rapidly from an experimental curiosity to an influential and sophisticated medium. The evolution illustrates how technology now affords creators an unprecedented degree of control over visual language, narrative structure, and viewer engagement (Wang, 2024). Generative tools, once constrained by early technical limitations that often rendered outputs as rudimentary or abstract, have now matured into advanced platforms capable of delivering polished, nuanced cinematic experiences (Liu et al., 2024). This transformation marks a pivotal moment in the history of visual media, one that positions the technology as a critical tool in reshaping creative expression.

Table 4.1 Core themes and reflections from leading creatives.

Theme	**Key Takeaways**	**Examples from Interviews**
Accessibility and Democratization	AI lowers barriers to entry, making high-quality production accessible to a wider range of creators. Artists can now create complex visuals without needing extensive resources or infrastructure.	Aashay Singh in India and Peter Zarko-Flynn in Australia use AI to overcome budget constraints, producing visually compelling narratives without costly traditional setups.
Transparency and Authenticity	Transparency around AI use is essential to maintain authenticity in creative works. Clear labeling and ethical guidelines are increasingly important as AI becomes more common in media.	Yonatan Dor labels his AI-generated content explicitly, aiming to prevent misunderstandings about AI's role in his work, especially in contexts involving high-profile or political figures.
Cultural and Regional Dimensions	Attitudes toward AI vary by region, influenced by local cultural and regulatory frameworks. These differences impact AI's application in creative fields, balancing innovation with ethical accountability.	Simon Graff notes Europe's cautious regulatory environment, whereas Singh in India highlights AI as an equalizer in regions with limited traditional filmmaking resources.
AI as Creative Partner vs. Efficiency Tool	AI serves as both a tool to streamline repetitive tasks and as a collaborative partner, enhancing the creative process without replacing human insight. Hybrid workflows blend traditional methods with AI's computational strengths.	Luka Tisler combines traditional VFX with AI for efficiency, while Christian Georgiev-Fries uses AI to enhance narrative and compositional control, seeing AI as a collaborator, not a replacement.
Interactivity and Audience Engagement	AI encourages interactive, participatory experiences, allowing audiences to engage dynamically with content. The shift to immersive media fosters new relational dynamics between creators and viewers.	Simon Graff's work in spatial computing and the Metaverse exemplifies how AI and immersive media merge to create responsive storytelling experiences.
Innovation and Experimentation	AI fosters experimentation, allowing artists to explore new formats, aesthetics, and narratives with reduced risk. This encourages creative risk-taking and aligns with early digital media's experimental spirit.	Aashay Singh's use of free tools like ChatGPT and Hugging Face to experiment with AI-driven film production reflects this innovative freedom.
Ethics and Intellectual Property	Navigating copyright and intellectual property remains a challenge in AI-driven art. Many artists prioritize responsible use by drawing from licensed or original content and maintaining ethical standards.	Peter Zarko-Flynn uses public domain works and original content, balancing creativity with respect for existing intellectual property.
Future Vision of AI in the Arts	Optimism about AI's potential to deepen audience engagement and personalization reflects a forward-looking vision. Many artists envision AI enhancing media's educational, immersive, and interactive dimensions.	Zarko-Flynn anticipates AI's role in democratizing film, while Singh sees potential for AI to be a tool for philosophical storytelling, and Graff envisions AI's integration with spatial computing for immersive media.

Table 4.1 contd. ...

...Table 4.1 contd.

Theme	Key Takeaways	Examples from Interviews
Hybridization of Traditional and Modern Techniques	AI enables a blend of traditional artistry with modern technology, bridging established practices with digital innovation. This approach maintains artistic integrity while expanding creative possibilities.	Luka Tisler's hybrid VFX-AI workflow and Christian Georgiev-Fries's use of AI to support traditional composition and form highlight the merging of traditional art principles with AI capabilities.
Educational and Historical Potential	AI's potential to preserve and reinterpret history opens avenues for educational applications, making complex or overlooked stories more accessible to audiences.	Zarko-Flynn's *Journey to Chernobyl* uses AI to animate still photographs from the exclusion zone, offering an emotional and immersive look into history through AI-driven storytelling.

The interviews collectively highlight the democratizing influence of AI on artistic production, a theme that parallels broader social and cultural trends towards inclusivity and access (Fui-Hoon Nah et al., 2023; Madan and Ashok, 2023). Given the web-based nature of the tools, artists from varied geographical and socio-economic backgrounds are now able to overcome traditional barriers associated with budget, location, and infrastructure (Gallus et al., 2023). For instance, Aashay Singh, working in India, and Peter Zarko-Flynn, based in Australia, both emphasize how such accessible tools have enabled them to produce visually intricate narratives without the need for high-cost resources typical of traditional filmmaking. Singh's projects, initially constrained by the lack of financial means for expansive cinematic endeavors, have benefited from the tools that allow for rapid, cost-effective experimentation with visual effects and narrative forms. Similarly, Zarko-Flynn's exploration of historical narratives, such as his piece *Journey to Chernobyl*, leverages video generators to animate still images and breathe life into archival material. The approach would have previously required extensive technical teams and resources, thus illustrating the capacity of the latest generation of intelligent systems to facilitate high-quality, impactful storytelling even for independent creators (Divya and Mirza, 2024).

These accounts align with the broader cultural movement observed in prior analyses: AI as a mechanism for accessibility in the arts (Hutson et al., 2023). Through the lowering of the threshold for participation, these new platforms have empowered a diverse range of voices to contribute to a medium that was once exclusive to those with institutional support or substantial funding. The work of Guillaume Hurbault in France and Yonatan Dor in Germany exemplifies this phenomenon, as both artists have utilized various AI-assisted platforms to explore innovative, visually compelling narratives that diverge from the conventions of mainstream media. Hurbault, lacking formal film training, has leveraged them to build a career in cinematic storytelling, producing projects that resonate with artistic depth and complexity. Dor, in turn, has pioneered such applications in the advertising and entertainment sectors, using generative tools to achieve high-caliber visual effects that rival traditionally produced media. These artists demonstrate how AI-driven media reconfigures the landscape of creative production, broadening the scope of participation and expanding the potential of visual storytelling.

The interviewed creators underscore essential considerations concerning the creative and ethical dimensions of AI, addressing themes of transparency, authenticity, and collaboration that resonate throughout the field (El Ali et al., 2024). For example, such concern was demonstrated in the case of Yonatan Dor of The Dor Brothers who clearly label their AI-generated videos released on social media, which underscores a rising awareness around authenticity, reflecting a broader cultural and ethical shift toward transparency. This aligns with discussions earlier in the book on the necessity for well-defined guidelines and ethical standards in AI art as these technologies become more pervasive (Darmantho, 2024; Taylor et al., 2024; Vyas, 2024). In the case of Dor, his stance highlights the importance of informing audiences, especially as deepfake technologies evolve and continues to blur the boundary between generated and authentic content, reinforcing the value of honesty in an era where this line grows increasingly ambiguous.

Another crucial insight from these creators pertains to the dual role of these platforms as both a creative partner and a tool for efficiency (Al Naqbi et al., 2024; McGuire et al., 2024). For many, the abilities of these generative tools are not seen as a replacement for traditional methods but as a collaborator that amplifies the creative process. Artists such as Luka Tisler and Christian Georgiev-Fries illustrate this nuanced relationship, where AI complements rather than replaces established practices. Tisler's hybrid approach—combining traditional VFX techniques with AI-generated components—demonstrates the practical benefits of blending human precision with computational strengths. Georgiev-Fries similarly embraces its ability to streamline technical tasks while emphasizing narrative and compositional integrity, reinforcing that the essence of visual storytelling is deeply rooted in human experience and emotion. The perspective echoes earlier themes in the book, challenging a simplistic binary view of AI as either threat or mere tool (Celis Bueno et al., 2024). Instead, these hybrid workflows suggest the potential for such tools to act as enhancer of artistic expression, allowing creators to engage more deeply with conceptual visions while leveraging AI to manage resource-intensive processes (Stige et al., 2023). Through these examples, the interviewed artists articulate a more nuanced view of the role played by emerging technologies, one that positions them as a supportive force in creative practice, thereby expanding possibilities in human-centered storytelling.

Another recurring theme throughout the interviews is the adaptability and experimentation that intelligent media technologies foster among artists, reflecting the iterative and flexible development that has marked creative automation historically. The progression of Aashay Singh from traditional to machine-driven film production, using accessible, free platforms like ChatGPT and Hugging Face servers, exemplifies how generative tools enable creators to explore new formats, styles, and aesthetics with minimal initial investment. Similarly, the educational initiatives led by Simon Graff in Germany highlight the role of synthetic media in sparking curiosity and adaptability among both seasoned professionals and newcomers. This trend underscores the flexibility and creative risk-taking that generative media encourages, echoing the experimental ethos that characterized the early rise of digital media in previous decades (Koivisto and Grassini, 2023).

At the same time, adoption is not a given and is driven by several factors. One such consideration is the cultural dimensions of automated media integration which are especially pronounced in terms of regional differences in adoption and regulation. The interviews reveal contrasting attitudes toward the potential of machine-generated content across diverse global contexts. For instance, Simon Graff notes the cautious regulatory landscape in Europe, where compliance concerns shape both corporate and creative applications of AI and require transparency in training data (Outeda, 2024). In contrast, Aashay Singh emphasizes the role of creative technologies as an equalizing force in India, where access to traditional filmmaking resources is often limited. This juxtaposition underscores a critical consideration for the future of intelligent media: as regulations evolve, there will likely be an ongoing need to balance innovation with ethical accountability, particularly in regions with varied technological infrastructure and levels of cultural readiness (Okem et al., 2023). Although the earlier chapters examined these regional dynamics, insights from the interviews add depth to this discussion, offering concrete examples of how regulatory and cultural environments shape the practical applications of generative media solutions.

Examining how these creatives use generative technologies to tell stories that might otherwise remain untold reveals the profound impact of AI on the evolving media landscape, echoing historical trends in digital democratization (Costa et al., 2024). Through their work, these artists demonstrate automated media's potential as both a visual and narrative tool, enabling creators to express complex, layered stories that resonate with diverse audiences. For instance, the goal of Luka Tisler of making high-quality VFX accessible to smaller markets and Peter Zarko-Flynn's immersive historical narratives underscore the role of these tools in supporting inclusive storytelling. The approach aligns with the overarching theme of the past chapters: technological evolution derives value not solely from novelty but from its capacity to enrich human experience, inviting more voices into the universal tradition of storytelling.

The interviews also reflect a forward-looking vision for the role of automated media in cinema and the visual arts. Each creator conveys optimism about the potential of generative tools to foster deeper, more interactive, and personalized audience engagement—an outlook that aligns with emerging trends across creative industries. The work of Simon Graff in spatial computing and the Metaverse suggests future possibilities for immersive media integration, while introspective narrative style of Aashay Singh highlights a growing interest in AI as a medium for philosophical and contemplative storytelling. These aspirations resonate with themes in earlier chapters, which explore the trajectory of generative media as a catalyst for richer, multidimensional interactions between creators and audiences. This forward-thinking perspective frames intelligent media as a tool that enhances and deepens the relational dynamics of storytelling, envisioning a future where technology amplifies, rather than diminishes, the authenticity of human narratives.

Synthesizing these insights reveals that integrating generative technologies into visual storytelling embodies both continuity and transformation within the arts, balancing traditional foundations with emerging technological possibilities. Core principles such as narrative structure, composition, and emotional resonance remain fundamental, anchoring creative works in a depth and intentionality that characterizes human-centered storytelling. Yet, generative tools contribute more than mere enhancement: they introduce unprecedented accessibility, enabling artists from diverse backgrounds to engage in high-caliber storytelling that once required substantial resources or institutional support. By lowering barriers to entry, these systems broaden the creative process, inviting a wider range of perspectives that enrich the cultural landscape with fresh interpretations of universal themes.

Moreover, machine-generated media fosters adaptability and innovation, transforming how artists approach storytelling in response to evolving audience expectations. Generative tools offer creators the flexibility to experiment, revise, and iterate swiftly, facilitating narratives that dynamically respond to audience engagement and contextual shifts. This adaptability is especially relevant in a media environment where interactivity and personalization are highly valued. As seen in the interviews, creators like Simon Graff envision applications within spatial computing and immersive environments that deepen audience interaction, making the storytelling experience more participatory. These advancements reflect automated media's role in redefining the relationship between creator and audience, fostering a co-creative environment that aligns with broader shifts toward engagement-driven media.

This duality—where generative tools serve as both facilitators of traditional practices and as disruptive forces—encourages a reexamination of creative norms. Artists such as Aashay Singh and Peter Zarko-Flynn view machine-driven media as a bridge connecting established artistic disciplines with cutting-edge technologies, expanding the boundaries of visual storytelling. By blending traditional cinematic techniques with generative systems, these artists explore uncharted realms of visual language, moving beyond conventional genres to tackle complex, layered narratives. This convergence exemplifies a movement that invites future generations to embrace creativity that bridges cultural, technological, and historical divides.

In this sense, synthetic media is not merely an augmentative tool but a catalyst for new forms of storytelling. It offers artists a framework to address diverse topics, from personal introspections to globally resonant social themes. The capability to generate visuals, emotions, and experiences that were once cost-prohibitive or logistically unfeasible enables more inclusive and innovative narrative expressions. As machine-driven media continues to evolve, it offers a pathway for bridging creative traditions with modern innovation, encouraging artists to explore new dimensions of storytelling that respond to technological advancements while rooted in enduring human experiences. This dynamic interplay between continuity and transformation underscores generative media's capacity to foster a future where storytelling transcends traditional constraints, inviting audiences and creators alike into a dialogue that bridges historical, cultural, and technological realms.

References

Al Naqbi, H., Bahroun, Z. and Ahmed, V. (2024). Enhancing work productivity through generative artificial intelligence: A comprehensive literature review. *Sustainability,* 16(3): 1166.

Cavdar, D. (2024). *Reimagining In/Visibility, Archival Practice and the Emergence of AI Cinema: An Exploration through Refik Anadol's Artworks* (Doctoral dissertation, Concordia University).

Celis Bueno, C., Chow, P. S. and Popowicz, A. (2024). Not "what", but "where is creativity?": towards a relational-materialist approach to generative AI. *AI & SOCIETY,* 1–13.

Chitrakorn, K. (2021). What influencer marketing looks like in the metaverse. *Vogue Business*. December 7, 2021: https://www.voguebusiness.com/technology/what-influencer-marketing-looks-like-in-the-metaverse.

Costa, C. J., Aparicio, M., Aparicio, S. and Aparicio, J. T. (2024). The democratization of artificial intelligence: theoretical framework. (18): 8236.

Darmantho, A. (2024). Copyright in the art industry: ethical and management challenges for artwork protection. *Jurnal Seni Musik,* 13(1): 42–58.

Divya, V. and Mirza, A. U. (2024). Transforming content creation: the influence of generative AI on a new frontier. *Exploring the Frontiers of Artificial Intelligence and Machine Learning Technologies,* 143.

Dueno, C. and Lopez-Figueroa, J. (2024). Fashion and AI: determining how the creative process for independent designers is shifting in 2023. *Journal of Student Research,* 13(2).

Durgade, B., Bakoda, Z., Kapuskar, R. and Salunkhe, S. (2024). Enhancing online shopping with AI avatars. *Zayed and Kapuskar, Rutuja and Salunkhe, Sanket, Enhancing Online Shopping with AI Avatars* (March 6, 2024).

El Ali, A., Venkatraj, K. P., Morosoli, S., Naudts, L., Helberger, N. and Cesar, P. (2024, May). Transparent AI disclosure obligations: who, what, when, where, why, how. pp. 1–11. *In*: *Extended Abstracts of the CHI Conference on Human Factors in Computing Systems.*

Evagelou, A., Kleftodimos, A. and Lappas, G. (2024). Creating location-based mobile applications for tourism: a virtual AR guide for Western Macedonia. *Digital,* 4(1): 271–301.

Faulconbridge, J., Sarwar, A. and Spring, M. (2023). How professionals adapt to artificial intelligence: The role of intertwined boundary work. *Journal of Management Studies.*

Frazen, C. (2024). WATCH: The Weekend's new concert teaser was made with Ai video, image generators. *Venture Beat.* August 23, 2024: https://venturebeat.com/ai/watch-the-weeknds-new-concert-teaser-was-made-with-ai-video-and-image-generators/.

Fui-Hoon Nah, F., Zheng, R., Cai, J., Siau, K. and Chen, L. (2023). Generative AI and ChatGPT: Applications, challenges, and AI-human collaboration. *Journal of Information Technology Case and Application Research,* 25(3): 277–304.

Gallus, P., Štěpánek, M., Ráčil, T. and Františ, P. (2023, October). Generative Neural Networks as a Tool for Web Applications Penetration Testing. pp. 1–5. *In*: *2023 Communication and Information Technologies (KIT)*. IEEE.

Gemünden, G. (2024). An Uncertain Future: The 74th Berlin Film Festival. *Film Criticism,* 48(1).

Hutson, J. (2024a, October 3). *Interview with Seth Steinacher* [Interview].

Hutson, J. (2024b, October 23). *Interview with Stephen Erin Dinehart IV* [Interview].

Hutson, J. (2024c, October 3). *Interview with Anup Gosavi* [Interview].

Hutson, J. (2024d, October 7). *Interview with PJ Accetturo* [Interview].

Hutson, J. (2024e, October 11). *Interview with Zahir Kahn* [Interview].

Hutson, J. (2024f, October 14). *Interview with Guillaume Hurbault* [Interview].

Hutson, J. (2024g, October 15). *Interview with Simon Graff* [Interview].

Hutson, J. (2024h, September 27). *Interview with Yonatan Dor* [Interview].

Hutson, J. (2024i, October 9). *Interview with Christian Georgiev-Fries* [Interview].

Hutson, J. (2024j, October 16). *Interview with Luca Tisler* [Interview].

Hutson, J. (2024k, October 9). *Interview with Aashay Singh* [Interview].

Hutson, J. (2024l, October 8). *Interview with Peter Zarko-Flynn (PZF)* [Interview].

Hutson, J., Lively, J., Robertson, B., Cotroneo, P. and Lang, M. (2023). *Creative Convergence: The AI Renaissance in Art and Design.* Springer Nature.

Izani, M., Razak, A., Rehad, D. and Rosli, M. (2024, August). The impact of artificial intelligence on animation filmmaking: tools, trends, and future implications. pp. 57–62. *In*: *2024 International Visualization, Informatics and Technology Conference (IVIT)*. IEEE.

Jääskeläinen, P., Holzapfel, A. and Eriksson, E. (2024, July). AI art for self-interest or common good? Uncovering value tensions in artists' imaginaries of AI technologies. pp. 2897–2910. *In*: *Proceedings of the 2024 ACM Designing Interactive Systems Conference.*

Jebb, L. (2024). Refik Anadol Studio reveals plans for world's first museum of AI arts. *The Art Newspaper.* September 24, 2024: https://www.theartnewspaper.com/2024/09/24/refik-anadol-studio-reveals-plans-for-worlds-first-museum-of-ai-arts.

Jetha, R. (2024). Watching the new AI-generated future of music videos. *The San Francisco Standard.* September 27, 2024: https://sfstandard.com/2024/09/27/ai-generated-music-videos-hackathon/.

Kart, J. (2020). Why IBM built an AI avatar to answer questions about marine litter. *Forbes*. August 10, 2020: https://www.forbes.com/sites/jeffkart/2020/08/10/why-ibm-built-an-ai-avatar-to-answer-questions-about-marine-litter/.

Kaur, G., Kaur, A. and Khurana, M. (2024, March). A survey of computational techniques for automated video creation and their evaluation. pp. 1–6. *In*: *2024 11th International Conference on Reliability, Infocom Technologies and Optimization (Trends and Future Directions) (ICRITO). IEEE.*

Kim, J. and Kim, H. (2024, May). Unlocking creator-AI synergy: challenges, requirements, and design opportunities in AI-powered short-form video production. pp. 1–23. *In*: *Proceedings of the CHI Conference on Human Factors in Computing Systems.*

Koivisto, M. and Grassini, S. (2023). Best humans still outperform artificial intelligence in a creative divergent thinking task. *Scientific Reports,* 13(1): 13601.

Kushnir, A. (2024). The new copyright frontiers–world-building and advanced technologies. pp. 179–200. *In*: *NFTs, Creativity and the Law.* Routledge.

Lafayette, J. (2024). TCL names finalists for AI TV/film accelerator program. *Next TV.* August 21, 2024: https://www.nexttv.com/news/tcl-names-finalists-for-ai-tvfilm-accelerator-.

Langely, H. (2020). Samsung's artificial humans aren't artificial or human enough. *Wired.* January 14, 2020: https://www.wired.com/story/samsung-neon-digital-avatars/.

Liu, Y., Zhang, K., Li, Y., Yan, Z., Gao, C., Chen, R. et al. (2024). Sora: A review on background, technology, limitations, and opportunities of large vision models. *arXiv preprint arXiv:2402.17177.*

Madan, R. and Ashok, M. (2023). AI adoption and diffusion in public administration: A systematic literature review and future research agenda. *Government Information Quarterly,* 40(1): 101774.

McCormack, J. and Grierson, M. (2024). Building simulations with generative artificial intelligence. pp. 137–150. *In*: *Climate Disaster Preparedness: Reimagining Extreme Events through Art and Technology*. Cham: Springer Nature Switzerland.

McGuire, J., De Cremer, D. and Van de Cruys, T. (2024). Establishing the importance of co-creation and self-efficacy in creative collaboration with artificial intelligence. *Scientific Reports,* 14(1): 18525.

NS. (2024). DaveAI's AI-Driven conversational interfaces transform financial services. APN News. July 10, 2024: https://www.apnnews.com/daveais-ai-driven-conversational-interfaces-transform-financial-services/.

Okem, E. S., Ukpoju, E. A., David, A. B. and Olurin, J. O. (2023). Advancing infrastructure in developing nations: a synthesis of AI integration strategies for smart pavement engineering. *Engineering Science & Technology Journal,* 4(6): 533–554.

Outeda, C. C. (2024). The EU's AI act: a framework for collaborative governance. *Internet of Things*, 101291.

Perelli, A. (2024). YouTube shorts feature leads to subscriber growth but revenue lags. *Business Insider*. February 11, 2024: https://www.businessinsider.com/youtube-shorts-leads-to-subscriber-growth-but-revenue-lags-2021-2.

Prance-Miles, L. (2020). Gucci develops partnership with Genies; users to create personalized avatars. *Global Cosmetics News*. October 23, 2020: https://www.globalcosmeticsnews.com/gucci-develops-partnership-with-genies-users-to-create-personalized-avatars/.

Reddy, V. S., Kathiravan, M. and Reddy, V. L. (2024). Revolutionizing animation: unleashing the power of artificial intelligence for cutting-edge visual effects in films. *Soft Computing,* 28(1): 749–763.

Rodgers, R. (2024). An avalanche of generative AI videos is coming to YouTube shorts. *Wired.* September 18, 2024: https://www.wired.com/story/generative-ai-tools-youtube-shorts-veo/.

Stige, Å., Zamani, E. D., Mikalef, P. and Zhu, Y. (2023). Artificial intelligence (AI) for user experience (UX) design: a systematic literature review and future research agenda. *Information Technology & People.*

Susskind, R. and Susskind, D. (2022). *The Future of the Professions: How Technology will Transform the Work of Human Experts.* Oxford University Press.

Tang, Y. and Chen, C. (2024). Can stylized products generated by AI better attract user attention? Using eye-tracking technology for research. *Applied Sciences,* 14(17): 7729.

Taylor, J., El Ardeliya, V. and Wolfson, J. (2024). Exploration of artificial intelligence in creative fields: generative art, music, and design. *International Journal of Cyber and IT Service Management,* 4(1): 39–45.

Vyas, B. (2024). Ethical implications of generative AI in art and the media. *International Journal for Multidisciplinary Research (IJFMR),* E-ISSN, 2582–2160.

Walk-Morris, T. (2024). Virtual influencers gain traction. *Marketing Dive.* September 26, 2024: https://www.marketingdive.com/news/virtual-influencers-gain-traction/728150/.

Wang, S. (2024, June). Research on interactive narrative design of immersive image from AI perspective. pp. 345–359. *In*: *International Conference on Human-Computer Interaction.* Cham: Springer Nature Switzerland.

Wingström, R., Hautala, J. and Lundman, R. (2024). Redefining creativity in the era of AI? Perspectives of computer scientists and new media artists. *Creativity Research Journal,* 36(2): 177–193.

Zhou, E. and Lee, D. (2024). Generative artificial intelligence, human creativity, and art. *PNAS Nexus,* 3(3): pgae052.

Chapter 5

Practical Guide to Workflows for AI Video Creation

This chapter provides readers with a structured guide for incorporating various AI video tools to streamline production workflows and enhance creative expression in video projects. It begins with a primer on creating AI art from text prompts using platforms such as MidJourney, emphasizing the nuances of crafting effective prompts and guiding users through initial setup to final output with illustrative examples and screenshots. The next section explains how to transform static visuals into animated sequences using tools like Pika and Runway, including detailed instructions and example videos that demonstrate the animation process. Following this, the integration of AI-generated visuals into traditional video editing environments shall be covered, detailing the combination of Adobe Premiere Pro and Cinema 4D for advanced post-production techniques, including quality enhancement, transitions, and effects. The chapter concludes with a discussion covering dialogue synthesis and lip-syncing with platforms such as Descript, Synthesia, and DeepBrain AI, as well as the addition of music through AI-powered scoring tools like AIVA and Soundraw. Together, these sections aim to empower creators to leverage AI-driven methods across the entire video production lifecycle, from ideation to polished multimedia output.

5.1 Creating AI Art from Text Prompts

Technology, particularly in the realm of generative AI, is advancing at a pace that outstrips Moore's Law, with new capabilities in image, video, and music generation tools appearing nearly every week (Jai and Shih, 2024). The rapid evolution of these models—each release pushing the boundaries of what each platform can produce—continually shifts the landscape, demanding adaptability from creators who work with these tools (Onyejelem and Aondover, 2024). Recognizing this pace of change, the following sections not only provide practical, step-by-step instructions for using the leading generative tools but also include best practices and broader considerations for the creative process. These larger guiding principles are designed to remain relevant as the specific features of these tools inevitably evolve, offering users a foundation to navigate the dynamic nature of AI-driven artistic workflows. With an understanding of the tools and techniques that underpin AI-generated visuals, creators can begin the multi-step process of assembling these individual elements into cohesive and immersive video narratives.

Generating video content with AI begins as a layered process where creators harness tools to craft each visual and auditory element of their work, shaping digital narratives from raw prompts to immersive experiences (Cho, 2024a). The journey often starts with foundational image generation using tools like MidJourney or DALL-E, where specific text prompts yield scenes, characters, or concepts (Ali et al., 2024). From these images, creators advance through animation stages: using image-to-video applications such as Runway or Pika, they transform still visuals into dynamic sequences that bring characters and settings to life. Text-to-video and video-to-video platforms extend this process, enabling creators to further build on these elements to shape cohesive scenes and evolving narratives, with each tool adding depth, motion, or stylistic variation to their AI-crafted universe (Fan et al., 2024).

As the sequences come together, the creative process continues into post-production, where visual effects, sound design, and editing refine the immersive quality of the project. At minimum, creators rely on professional editing software, such as Adobe Premiere Pro, to piece together individual clips and establish continuity through refined transitions, pacing, and style (Rahman and Ali, 2024). In this stage, creators address the unique challenges of AI-generated media: adjusting the lip-syncing of virtual actors, aligning facial expressions with scripted dialogue, and adding sound effects that enhance realism. Platforms like Descript and Synthesia offer advanced lip-syncing, while AIVA and Soundraw supply scoring options that strengthen the emotional resonance of each scene (Alshahrani and Maashi, 2024). By the end of this evolving process, the combined power of visual and audio AI tools transforms initial prompts into a richly layered narrative, echoing the artistic vision behind each generative decision.

The initial step in AI-driven video creation often begins with generating the visual components using AI art generators. These tools enable creators to rapidly produce a foundation of digital images based on descriptive text prompts, opening up endless possibilities for scene-setting, character design, and conceptual art. Each AI generator brings unique features and strengths to the creative process: some offer highly customizable outputs, while others focus on stylistic interpretations that expand upon concise prompts. In this section, we examine four of the most popular AI art generators—Stable Diffusion, DALL·E 3, Midjourney, and Adobe Firefly—each offering distinctive advantages that cater to diverse creative objectives. By understanding the unique capabilities of these platforms, artists can strategically incorporate them into their workflows, crafting initial images that serve as the building blocks for animation, storytelling, and immersive visual narratives.

Stable Diffusion, an open-source model released in 2022, is renowned for its adaptability and customizability. Built as a deep learning model, it excels in producing highly detailed images from descriptive prompts and supports tasks such as inpainting (filling in missing parts of an image), outpainting (expanding an image beyond its original boundaries), and image-to-image translations guided by a text prompt. Stable Diffusion's distinctive feature lies in its underlying architecture: it operates on a latent diffusion model, a deep generative neural network that streamlines the process of image generation by iteratively removing Gaussian noise. This design allows for efficient training and generation while retaining intricate details. Stable Diffusion's open-source code and accessible model weights enable users to run it on consumer hardware with a standard GPU, making it one of the more versatile tools for creators who seek full control over the final output. However, its extensive options require a higher level of technical skill, making it ideal for users familiar with prompt engineering and model customization.

DALL·E 3, OpenAI's latest image generator, excels at converting detailed textual descriptions into high-quality visuals with remarkable interpretative ability. Integrated into ChatGPT's interface, DALL·E 3 facilitates a unique, conversational approach to crafting prompts, allowing users to iteratively refine descriptions to yield highly relevant and personalized outputs. This tool is especially adept at handling intricate scenes and generating visuals with complex compositions, including textual elements, making it suitable for applications that demand high levels of precision and illustrative detail. As the latest evolution of the original DALL-E model, DALL·E 3 maintains the capability to combine disparate concepts, styles, and attributes, though its image resolution remains slightly lower compared to models like Stable Diffusion and MidJourney.

DALL·E 3 operates with a multimodal framework and is accessible to pro ChatGPT subscribers. Unlike its predecessor, DALL·E 2, which offered four images per prompt for users to select from, DALL·E 3 generates images sequentially, providing one output at a time with an hourly limit on the number of images that can be created. This change aligns with OpenAI's intention to optimize the image quality and focus on creating a more interactive user experience through conversational prompt modification. The anticipated release of ChatGPT 5 promises to enhance this workflow by integrating video generation capabilities directly into the platform, eliminating the need to transition from static image creation to video models. This evolution would streamline the creative process, enabling users to generate fully realized visual and video content within a single platform, furthering OpenAI's goal of building an all-encompassing creative AI suite.

Midjourney, an AI art generator developed by Midjourney, Inc., has gained a dedicated following among digital artists and video creators for its distinct, imaginative aesthetic. Known for its responsiveness to concise, open-ended prompts, Midjourney often interprets text inputs with a level of artistic intuition that surprises users, yielding uniquely stylized and visually compelling results. This capacity for interpretative creativity makes Midjourney particularly popular among creators aiming for conceptual or abstract visuals, where the tool's inherent artistry adds depth and originality to the final output. As a result, it has become a go-to resource for projects where mood, style, and visual narrative play a central role.

Initially, users accessed Midjourney via Discord, interacting with the model by submitting commands and prompts to a bot that generated artwork in response. While powerful, this approach came with limitations; many users found the interface cumbersome and its command-based structure less intuitive for visual workflows.

Recognizing the need for a more user-friendly experience, Midjourney, Inc. has since introduced a dedicated webpage with an improved user interface and enhanced user experience elements, enabling easier access and more streamlined interaction with the model. This evolution has expanded Midjourney's accessibility, allowing artists to navigate their creative process with greater ease and focus on leveraging the model's aesthetic strengths. The shift to a web-based platform positions Midjourney as an adaptable tool, one that continues to balance technical sophistication with an intuitive user experience tailored for creators of all backgrounds.

Adobe Firefly, a specialized family of AI models, is tailored for seamless integration within Adobe's suite of creative applications, particularly benefiting users of Photoshop, Illustrator, and other Creative Cloud services. Firefly's unique appeal lies in its design focus: it generates commercially safe content by training on Adobe Stock, public domain images, and licensed content, which means creators can work without concerns about intellectual property issues. Firefly offers features such as text-to-image generation, text effects, and generative fill, aimed at enhancing creative workflows through AI-powered automation. This design allows users to ideate, produce, and manipulate images and text seamlessly within Adobe's ecosystem, making Firefly a valuable addition for users already embedded in the Adobe environment.

Firefly's recent expansion integrates it directly into Adobe's core services, including Creative Cloud, Document Cloud, Experience Cloud, and Adobe Express, under the broader Adobe Sensei generative AI services. Despite its user-friendly integration, Firefly's training data limitations—focusing on licensed and out-of-copyright materials—can sometimes result in outputs that lack the realism and nuanced control seen in other models like Stable Diffusion or Midjourney. Many users note that this constrained dataset impacts the quality of generated visuals, particularly in areas requiring fine details or complex textural work. However, Adobe has aimed to balance these limitations by incorporating generative video capabilities within Firefly, positioning it as a comprehensive tool for creators who value workflow convenience. This addition allows users to move from static image generation to dynamic video creation within the same environment, enhancing Firefly's utility and solidifying its role in Adobe's expansive creative ecosystem.

Building on the distinct strengths of these four AI image generators, understanding the nuances of effective prompting is essential for maximizing the creative potential of each tool. Effective prompt engineering for image generation models such as Stable Diffusion, DALL·E 3, Midjourney, and Adobe Firefly involves an understanding of best practices that apply across these platforms. While each model interprets language in unique ways, there are overarching strategies that enhance the quality and relevance of generated images. One key consideration is that these images often serve as foundational elements for video production; thus, prompts should be crafted with a cinematic vision in mind. This means describing not only the subject or scene but also specifying details such as lighting conditions, camera angles, and composition. Prompting for ambient lighting, for example, can produce subtle shadow work, while mentioning "overhead angle" or "close-up shot" will set up the scene's visual tone, making it more effective as a video input.

It is also crucial to understand that prompting does not have one "correct" methodology. Platforms like YouTube now offer thousands of hours of training on prompt techniques, shared by various creatives who bring distinct approaches. This variety demonstrates that while there are common practices, prompt engineering is also an art, and users are encouraged to experiment. As AI models evolve, they are increasingly adept at handling more generalized prompts without needing the meticulous detail required in earlier versions. While past models demanded carefully constructed prompts to achieve quality outputs, newer, larger models can often intuitively interpret shorter or broader descriptions, allowing users more flexibility in their approach. However, for those seeking specific visual attributes in their image outputs, guiding the model with targeted keywords remains a valuable skill (Sahoo et al., 2024).

While Stable Diffusion, DALL-E 3, Midjourney, and Adobe Firefly each leverage AI to transform textual prompts into visual art, their unique attributes offer creators a toolkit of diverse possibilities suited to various artistic goals. From the precision and flexibility of Stable Diffusion to the detailed synthesis in DALL-E 3, the imaginative outputs of Midjourney, and the seamless Adobe integration in Firefly, each tool allows artist to push creative boundaries in distinct ways. The adaptability of these models means that creators can strategically choose the tool that best aligns with their vision, whether that involves intricate detail, stylistic experimentation, or streamlined workflow integration. As such, the following sections provide detailed, step-by-step instructions to guide users through each platform, demonstrating how to effectively employ these tools to achieve desired results. This hands-on approach will illustrate how various platforms can enhance creativity and help filmmakers achieve their envisioned aesthetic more accurately.

5.1.1 DALL-E 3

DALL·E 3, introduced by OpenAI in September 2023, represents a significant advancement in AI-driven image generation. Building upon its predecessors, DALL·E and DALL·E 2, this latest iteration offers enhanced capabilities in translating complex textual descriptions into high-quality, detailed images. A notable feature of DALL·E 3 is its integration with ChatGPT, allowing users to generate images through conversational interactions, thereby streamlining the creative process. Because of such integration, access to DALL·E 3 is available to ChatGPT Plus and Enterprise users (openai.com), with a subscription fee of $20 per month. This subscription provides unlimited access to both ChatGPT and DALL·E 3, enabling users to generate images directly within the ChatGPT interface. Additionally, DALL·E 3 is accessible through Microsoft's Bing Chat, broadening its availability to a wider audience.

Technically, DALL·E 3 is built upon the GPT-4 architecture, leveraging its advanced language understanding to interpret and generate images from detailed prompts. This integration allows for more accurate and nuanced image generation, effectively capturing intricate details and complex scenes. The model has been trained on a diverse dataset of text-image pairs, enabling it to produce images across various styles, including photorealistic imagery, paintings, and illustrations.

At the same time, it also incorporates safety measures to prevent the generation of harmful or inappropriate content. The model is designed to decline requests that involve explicit, violent, or hateful imagery, as well as prompts that may infringe on copyright or depict public figures without consent. While these safeguards are part of OpenAI's commitment to responsible AI deployment, creatives need also be aware of these limitations when selecting a model that can generate the subject matter desired. Nevertheless, the model remains a powerful tool for ideation and inspiration. Its ability to generate diverse and detailed images from textual descriptions opens new avenues for creative exploration, concept development, and visual storytelling. Whether used for brainstorming, prototyping, or artistic experimentation, DALL·E 3 provides a valuable resource for expanding creative possibilities. Step-by-step instructions for using the model can be found in Table 5.1.

Using DALL-E 3 to generate images offers creators an intuitive way to translate detailed prompts into vivid visuals. To maximize the model's potential, especially when creating assets intended for video generation, users should consider specific techniques in prompt crafting and image refinement. Once signed up for ChatGPT Plus and logged into the ChatGPT platform, start by selecting the GPT-4 model to access DALL-E 3. Here, you can engage directly in conversational prompt creation, refining image outputs in real-time based on feedback (Fig. 5.1). This interactive approach enables you to provide iterative adjustments, helping the AI understand the desired scene, composition, and style with greater precision.

When generating images for video, it is essential to frame prompts that specify core visual attributes such as lighting, angles, and stylistic details. For instance, if designing a forest scene for a video, you might start with: "Create a dense forest at dusk, with mist rising between trees and a soft, warm light filtering through the branches" (Fig. 5.2). Including lighting descriptors like "soft" or "dusk" can add mood, while specifying angles—such as "overhead" or "low angle"—ensures consistency in perspective when these assets are later combined into video sequences

Table 5.1 Guide to generating images using DALLE-3.

Step	Description
1. Set Up an OpenAI Account	Visit OpenAI's website and sign up for an account. This will grant you access to OpenAI's services, including ChatGPT and DALL·E 3.
2. Subscribe to ChatGPT Plus	To access DALL·E 3, subscribe to ChatGPT Plus for $20 per month. This subscription provides enhanced features and priority access to new tools.
3. Access ChatGPT	Log in to your OpenAI account and navigate to ChatGPT. Ensure you're using the GPT-4 model, which integrates DALL·E 3 capabilities.
4. Input Prompts to Generate Images	In the chat interface, type a detailed description of the image you want to create. For example: "Create an image of a serene mountain landscape at sunrise." DALL·E 3 will generate an image based on your prompt.
5. Revise Images as Needed	If the generated image isn't to your liking, provide additional instructions or modify your prompt to refine the image. You can request changes such as adjusting colors, adding elements, or changing styles.
6. Download the Final Image	Once satisfied, click on the image to view it in full size. Use the download option to save the image to your device.

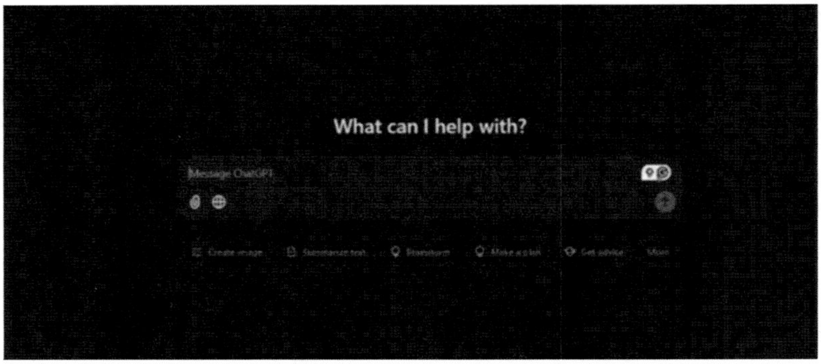

Figure 5.1 ChatGPT Interface, OpenAI, 2024.

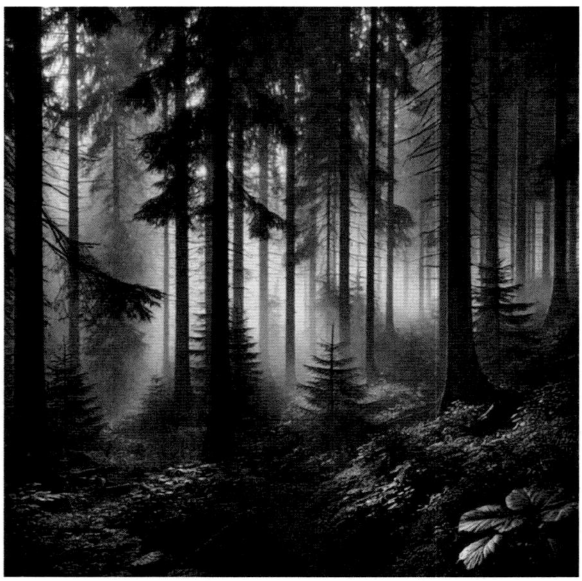

Figure 5.2 'Dense Forest at Dusk' prompt, DALLE-3, OpenAI, 2024. Courtesy of authors.

(Fig. 5.3). Moreover, mention textures or environmental details that enhance depth and realism; these elements add continuity when moving from static images to animated sequences. With DALL-E 3's ChatGPT integration, refining prompts is straightforward. If an image lacks particular visual details, prompt the model with suggestions such as "add more mist in the background" or "enhance the shadows." This iterative feedback process ensures that images are optimized for realism and detail, essential for later video integration. Adjusting color schemes and contrast through refined prompts also helps maintain visual cohesion when transitioning from one image to another in video form.

Once you have a refined set of images, review them for consistency in lighting, angle, and detail. Videos benefit from visual uniformity across frames, so ensure that images share similar attributes. Higher consistency here reduces the workload

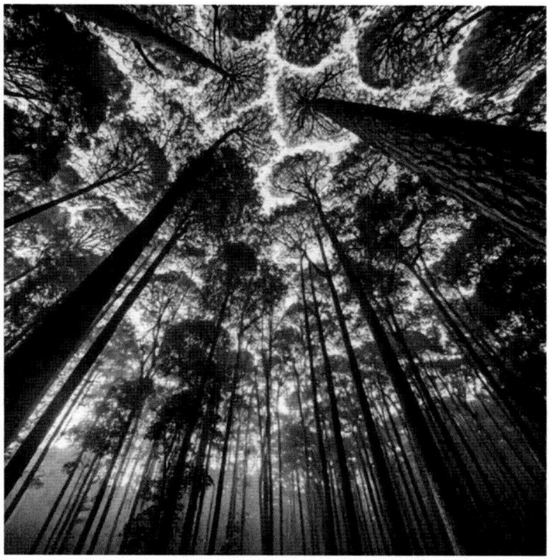

Figure 5.3 'Dense Forest at Dusk from Low Angle' prompt, DALLE-3, OpenAI, 2024. Courtesy of authors.

in video editing, creating a more fluid visual narrative. Finally, save the images in high resolution, as quality loss can impact video clarity when using video generators. Using images that are prepared with detailed prompts and iterative refinement in DALL-E 3 ensures that your visuals will translate smoothly from static images to dynamic video.

5.1.2 Midjourney

Midjourney (https://www.midjourney.com) is a generative AI platform developed by Midjourney, Inc., an independent research lab based in San Francisco, California. Led by David Holz, co-founder of Leap Motion, the team launched Midjourney into open beta on July 12, 2022. The platform enables users to create images from natural language descriptions, known as "prompts," similar to OpenAI's DALL-E and Stability AI's Stable Diffusion. Unlike some AI tools that require detailed prompts, Midjourney encourages users to provide concise and general prompts, allowing the AI to exercise its creative interpretation (Tsidylo and Sena, 2023). This approach fosters a collaborative relationship between the user and the AI, resulting in visually intriguing and diverse images based on minimal input. By embracing simplicity in prompts, Midjourney explores various creative directions, producing unconventional outputs that may not be predictable from the initial prompt alone.

To use Midjourney, users have the option to interact with a Discord bot by entering the command "/imagine" followed by their prompt. The bot then generates a set of images based on the description provided. This process allows artists, designers, and creators to explore new realms of visual expression, spark inspiration, and discover unexpected visual possibilities. The other option to interact with the model is via web. Midjourney has expanded its accessibility by introducing a web-based interface, allowing users to generate images directly through their browser without relying on Discord (Cho, 2024b). This development offers a more user-friendly experience, enabling artists, designers, and creators to craft AI-generated visuals seamlessly. The web platform consolidates various features, including inpainting, outpainting, and image resizing, into a unified editor, streamlining the creative process. By eliminating the need for Discord, Midjourney broadens its reach, making AI-driven art creation more accessible to a diverse audience.

The unique approach and continuous development of Midjourney offer a powerful AI tool for those looking to delve into generative art and unleash their creative potential. Whether used as a standalone tool or in combination with other AI-driven or traditional creative methods, Midjourney provides an innovative platform for generating artistic content. A step-by-step guide on using the tool can be found in Table 5.2.

Creating images with Midjourney offers a dynamic experience where users can guide the creative direction of the model with minimal input, ideal for those looking to produce imaginative visuals with an open-ended approach. With options for both a Discord-based and a newly launched web interface, Midjourney is accessible

Table 5.2 Guide to generating images using Midjourney.

Step	**Discord Interface**	**Web Interface**
1. Set Up an Account	Create or log in to your Discord account.	Visit Midjourney's website and log in using your Discord credentials.
2. Join Midjourney	Navigate to the Midjourney Discord server and accept the invite to join.	Access the web interface directly after logging in.
3. Subscribe to a Plan	In the Discord server, use the /subscribe command to receive a link to the subscription page, then choose a plan that fits your needs.	On the website, navigate to the subscription section and select a suitable plan.
4. Access the Bot or Interface	In Discord, locate a channel labeled #newbies-# or a direct message with the Midjourney Bot.	On the website, go to the "Create" tab to start generating images.
5. Generate an Image	Type /imagine followed by your prompt (e.g., /imagine a serene mountain landscape at sunrise) and press Enter.	Use the "Imagine" bar at the top of the page to enter your prompt and press Enter.
6. Review and Refine	The bot will generate four image variations. Use the U buttons to upscale or the V buttons to create variations.	The web interface will display the generated images with options to upscale, create variations, or edit further.
7. Download the Image	Click on the upscaled image to view it in full size, then right-click and select "Save Image As" to download.	Click on the desired image to open it, then use the download option provided to save it to your device.

and versatile, empowering artists to experiment with AI-generated art quickly and efficiently. The web interface, in particular, simplifies the process, allowing users to enter prompts, view results, and refine images all within one cohesive platform.

When generating images in Midjourney, the key to maximizing output quality is in crafting prompts that strike a balance between specificity and openness. Unlike certain other AI tools that benefit from highly detailed prompts, Midjourney thrives on brevity. Prompts like "sunlit forest path at dawn" (Fig. 5.4) or "surreal dreamscape with floating islands" allow the model to exercise its interpretive power, filling in visual gaps to create nuanced and often unexpected outcomes. However, if the goal is to prepare images for video generation, incorporating specific stylistic cues—such as "soft lighting," "cinematic angle," or "high contrast"—ensures continuity across frames, which is essential when assembling visuals into a cohesive video (Fig. 5.5).

An additional best practice is to use Midjourney's built-in options for upscaling and creating variations, available through either the Discord bot or the web interface. Upscaling improves image resolution, crucial for clarity when integrating images into video sequences. Creating variations of a single prompt can yield multiple angles or interpretations of the same scene, which is valuable for crafting different shots or perspectives in video production. Once satisfied with the generated images, saving them at the highest available resolution ensures optimal quality for further use in video generators. By following these practices, artists can seamlessly transition Midjourney's imaginative outputs into high-quality assets for video projects, leveraging the model's unique creative capacity to produce visuals that are both compelling and video-ready.

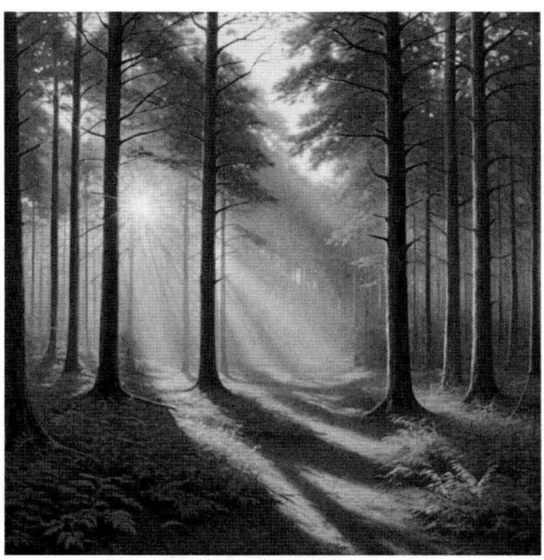

Figure 5.4 'Sunlit Forest Path at Dawn' prompt, Midjourney, 2024. Courtesy of authors.

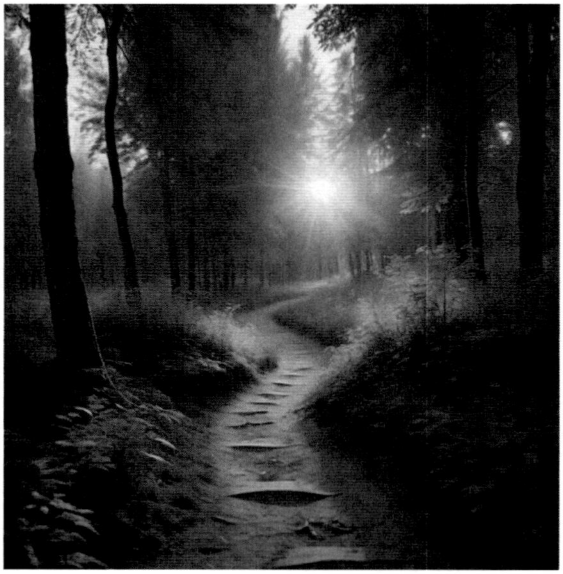

Figure 5.5 'Sunlit Forest Path at Dawn with Soft Lighting at a Cinematic Angle' prompt, Midjourney, 2024. Courtesy of authors.

5.1.3 Stable Diffusion

Stable Diffusion (https://stablediffusionweb.com/), a deep learning text-to-image model developed by Stability AI, has rapidly evolved since its initial release in 2022. Created in collaboration with academic researchers and non-profit organizations, Stable Diffusion enables users to generate detailed images based on textual descriptions. The model utilizes a latent diffusion architecture composed of a variational autoencoder (VAE), a U-Net, and a text encoder, creating a sophisticated pathway for producing high-quality images. During the diffusion process, Gaussian noise is added to images in a compressed latent space, which is then progressively removed by the U-Net before final reconstruction by the VAE. This structure allows Stable Diffusion to produce vivid images with intricate detail, and the text encoder (CLIP ViT-L/14) enables users to shape these visuals through natural language prompts. Its extensive training on the LAION-5B dataset, a large corpus of image-caption pairs filtered by language, resolution, watermark likelihood, and aesthetic quality, allows Stable Diffusion to generate a wide range of diverse and compelling imagery.

The latest model, Stable Diffusion 3.5, introduced in October 2024, brought several enhancements to the platform. Stable Diffusion 3.5 comes in three versions: a large model with 8.1 billion parameters for high-resolution, professional-grade imagery; a turbo variant for rapid generation with lower computational demand; and a medium version, with 2.5 billion parameters, designed to work effectively on consumer-grade hardware (Dehouche and Dehouche, 2023). These variants enable Stable Diffusion to cater to both professional and hobbyist users, making high-quality generative art accessible across different hardware configurations. The inclusion of

Query-Key Normalization in the model's architecture further stabilizes training and allows users to fine-tune images according to specific creative needs. Additionally, improvements to image diversity and style adaptability make it easier for users to create varied outputs without needing excessively detailed prompts.

The progressive diffusion approach of this model has earned it a reputation for creating high-resolution images with sharp details, smooth transitions, and intricate textures. These qualities make it particularly suitable for applications in digital art, graphic design, and computer-generated imagery, where precision and fidelity are essential. Since it is open-source, Stable Diffusion can be run on consumer hardware with a standard GPU, democratizing access to high-quality image generation technology. Platforms like OpenArt and Diffus.me have introduced user-friendly interfaces that further enhance accessibility, complete with tutorials to guide users through the process. Overall, Stable Diffusion represents a significant leap forward in text-to-image generation. Its combination of flexibility, high fidelity, and open-source availability sets it apart from proprietary models, empowering a diverse community of artists, designers, and researchers to explore new creative possibilities in AI-generated art. A step-by-step approach to generating images with the model can be seen in Table 5.3.

Using Stable Diffusion to create images provides creators with a powerful and versatile tool to generate high-resolution visuals through text prompts. As a deep learning model, Stable Diffusion thrives on detailed descriptions, offering users fine control over the style, texture, and lighting of their images. To begin, creators access Stable Diffusion through platforms like DreamStudio or, for more advanced users, install it locally on compatible hardware. The process starts by crafting a carefully

Table 5.3 Guide to generating images using Stable Diffusion.

Step	Description
1. Set Up an Account	Visit Stability AI's website and sign up for an account to access Stable Diffusion.
2. Choose a Platform	Decide whether to use Stable Diffusion via a web-based platform like DreamStudio or install it locally on your computer.
3. Access the Platform	If using DreamStudio, log in with your Stability AI credentials. For local installation, follow the setup instructions provided by Stability AI.
4. Input a Text Prompt	Enter a descriptive text prompt detailing the image you wish to generate. For example: "A serene mountain landscape at sunrise."
5. Configure Settings	Adjust parameters such as image resolution, number of inference steps, and guidance scale to refine the output. Higher inference steps can lead to more detailed images but may increase processing time. OpenArt
6. Generate the Image	Click the 'Generate' button to initiate the image creation process. The system will process your prompt and settings to produce the image.
7. Review and Refine	Examine the generated image. If it doesn't meet your expectations, modify the prompt or settings and regenerate until satisfied.
8. Download the Image	Once content with the result, download the image to your device for use in your projects.

considered text prompt that captures the essential details of the desired image—whether it's a scene, character, or abstract concept.

For users intending to integrate these images into video, prompt specificity is particularly critical. Stable Diffusion performs best when given detailed prompts that include lighting conditions (e.g., "soft morning light"), perspective (e.g., "overhead angle"), and texture (e.g., "foggy and misty") to ensure continuity across frames. For instance, the following would be an appropriate prompt to get started: "A misty forest pathway at dawn, with soft, diffused sunlight filtering through tall trees, casting gentle shadows across the ground. The scene has a cinematic feel with a wide-angle perspective, intricate leaf textures, and subtle fog layering in the distance. Colors are muted and natural, with hints of morning light illuminating the mist, creating a serene and dreamlike atmosphere" (Fig. 5.6). Through the inclusion of these elements, creators establish a cohesive aesthetic that translates smoothly into video form. Once the prompt is entered, Stable Diffusion's configurable settings, such as image resolution, inference steps, and guidance scale, provide additional control over the visual output. Increasing the inference steps, for example, can yield finer details, enhancing clarity in close-up shots.

After generating an image, reviewing and refining the output is essential. Stable Diffusion allows users to make iterative adjustments by tweaking prompts or altering settings, producing refined visuals that are ideal for video integration. Once satisfied, users can download the images in high resolution, preserving quality for video upscaling and editing. This preparation is crucial, as low-resolution images can result in pixelation and loss of detail during animation. Additionally, maintaining a consistent color palette and style across images simplifies the video assembly process, reducing the need for extensive post-production adjustments. By applying these best practices, creators can effectively use Stable Diffusion to generate high-quality visuals tailored for seamless video integration. The model's adaptability to detailed prompts and configurable settings makes it a valuable asset in producing imagery that aligns with specific video project needs, enhancing the overall visual coherence and impact of the final production.

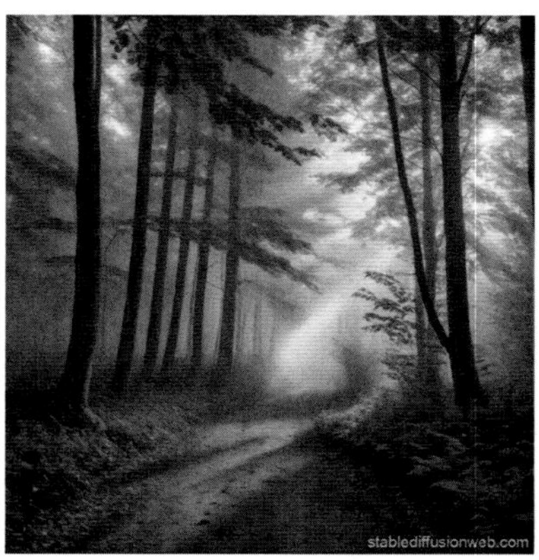

Figure 5.6 'A Misty Forest Pathway at Dawn...' prompt, Stable Diffusion, 2024. Courtesy of authors.

5.1.4 Adobe Firefly

Adobe Firefly (https://www.adobe.com) is a generative AI suite specifically designed to enhance the creative workflows within Adobe's ecosystem. Since its initial unveiling on March 21, 2023, Firefly has been integrated across Adobe's major platforms, including Creative Cloud, Document Cloud, Experience Cloud, and Adobe Express. This integration allows users to access Firefly's generative AI capabilities directly within applications like Photoshop, Illustrator, and Premiere Pro, making it easier for professionals and creators to incorporate AI-driven media elements seamlessly into their projects. By embedding these tools into widely-used Adobe applications, Firefly brings generative AI capabilities closer to the workflows of designers, marketers, and artists, streamlining the creative process and enabling rapid experimentation and ideation.

Adobe has prioritized ethical considerations in the development of Firefly by training its models exclusively on content that is either licensed or in the public domain. This approach not only addresses potential copyright concerns but also ensures that the generative AI respects intellectual property rights within the creative community. Adobe has further committed to introducing a "Do Not Train" tag, allowing creators to exclude their content from future training datasets, thus enabling artists to have control over how their work is used in AI model development. These safeguards make Firefly one of the few AI models that offer robust ethical practices, aiming to set a standard in responsible AI deployment (Angelova, 2024).

Since its beta release, Adobe Firefly has expanded significantly in terms of features and capabilities. Among its most popular tools are Generative Fill and Generative Expand, which allow users to add or extend visual content within images, providing greater flexibility in creative exploration. Additionally, Firefly's text-to-image and text effects functionalities enable users to generate entire images or stylized text by simply inputting descriptive prompts. This feature is especially valuable for rapid concept visualization and design iteration. Recently, Adobe has extended Firefly's reach into video content creation, allowing users to generate video clips based on text prompts or still images. This expansion into video generation has made Firefly a versatile tool, suitable for a wider range of creative applications beyond static imagery (Huang, 2024).

The approach of Adobe to Firefly development has been collaborative, involving regular feedback from the creative community. This ongoing dialogue has played a crucial role in refining the AI models and adding new features that align with the needs of creative professionals. Adobe's commitment to engaging with users ensures that Firefly continues to evolve in ways that enhance its usability and adaptability across various creative fields. By continually updating the model with user-informed improvements, Firefly is positioned to stay at the forefront of AI tools for creative professionals, meeting industry standards and addressing real-world challenges.

To further support creators, Adobe is developing monetization options for users of Firefly-generated content. These initiatives include compensation models for Adobe Stock contributors, allowing artists to benefit financially from their work within Firefly's ecosystem. This strategy fosters a sustainable creative economy, where professionals not only create with AI but also gain opportunities for income, thus reinforcing Adobe's commitment to supporting a thriving and ethically-conscious creative community. In sum, Adobe Firefly represents a well-rounded approach to generative AI, combining innovative tools, ethical practices, and community engagement to empower users while addressing the complex needs of modern content creation. A guide to generating images using the interface can be seen in Table 5.4.

Creating images with Adobe Firefly provides users a streamlined experience that leverages Adobe's robust design ecosystem, enabling high-quality image generation through intuitive tools. The process begins with accessing the Firefly platform, where users can select the 'Text to Image' option and input prompts that guide the AI in generating visuals based on descriptive language. For best results, prompts should be detailed and purposeful, specifying scene elements, color tones, lighting conditions, and style preferences. Firefly's distinct advantage is its tight integration

Table 5.4 Guide to generating images using Adobe Firefly.

Step	**Description**
1. Access Adobe Firefly	Navigate to the Adobe Firefly website and sign in with your Adobe ID. If you don't have an account, you can create one for free.
2. Select 'Text to Image'	On the Firefly homepage, click on the 'Text to Image' option to begin creating images from text prompts.
3. Enter Your Prompt	In the prompt field, type a detailed description of the image you want to generate. For example: "A serene mountain landscape at sunrise with a clear blue sky." Adobe Help Center
4. Adjust Settings	Utilize the settings panel to refine your image. Options include:
	- **Aspect Ratio:** Choose the desired dimensions for your image.
	- **Content Type:** Select 'Art' or 'Photo' to define the style.
	- **Styles:** Apply artistic styles or effects to your image.
	- **Color and Tone:** Adjust the color palette and mood.
	- **Lighting:** Modify lighting conditions to suit your vision.
	- **Composition:** Influence the arrangement of elements within the image.
5. Generate the Image	After configuring your settings, click the 'Generate' button. Firefly will process your prompt and display four image variations based on your description. Adobe Help Center
6. Review and Refine	Examine the generated images. If they don't meet your expectations, you can:
	- **Edit the Prompt:** Modify your description for different results.
	- **Adjust Settings:** Change styles, lighting, or other parameters.
	- **Regenerate:** Click 'Generate' again to produce new variations.
7. Download Your Image	Once satisfied, select the desired image and click the 'Download' icon to save it to your device in JPEG or PNG format. Adobe

with Adobe Creative Cloud, making it an ideal choice for creators aiming to develop assets with consistency and cohesion.

When crafting prompts for images that will eventually be uploaded to video generators, users benefit from considering factors that enhance visual continuity across scenes. Descriptions like "warm sunrise lighting over a serene mountain landscape" or "wide-angle shot with soft, diffused light" help set consistent color tones and lighting, attributes crucial for scenes meant to flow together in a video sequence (Fig. 5.7). Firefly also provides specific adjustments for artistic style, aspect ratio, color palette, and composition, allowing users to refine their images further. Experimenting with these options can yield stylistically cohesive images that require minimal adjustments when assembled into a video.

Once an image is generated, reviewing and refining it through Firefly's settings is key. Users can make iterative adjustments to lighting, composition, or specific styles, ensuring that each asset aligns with the intended visual narrative. Firefly's four-image generation option offers multiple interpretations of the same prompt, giving users a chance to explore variations and select the best fit. After refining the images, users can save them in high resolution, ready for seamless integration into video projects where quality and detail are essential. These best practices allow creators can maximize the capabilities of the model, producing high-quality visuals optimized for cohesive and impactful video storytelling.

These popular tools are only a sampling of the broader landscape of generative AI platforms. Each model introduced here offers unique features, yet numerous other tools (many of which are free) cater to specific needs or niches within the field, expanding the creative options available to artists, designers, and content creators. For instance, NightCafe (https://creator.nightcafe.studio/), known for its diverse algorithmic capabilities, allows users to generate original art from text prompts and features a robust, community-driven platform with a credit-based system. This model emphasizes user interaction and sharing, making it ideal for creators who wish to explore various algorithms and styles in a collaborative environment (Limano, 2023). Similarly, DeepAI (https://deepai.org/machine-learning-model/text2img) offers a comprehensive suite of AI tools for generating and enhancing images, allowing users

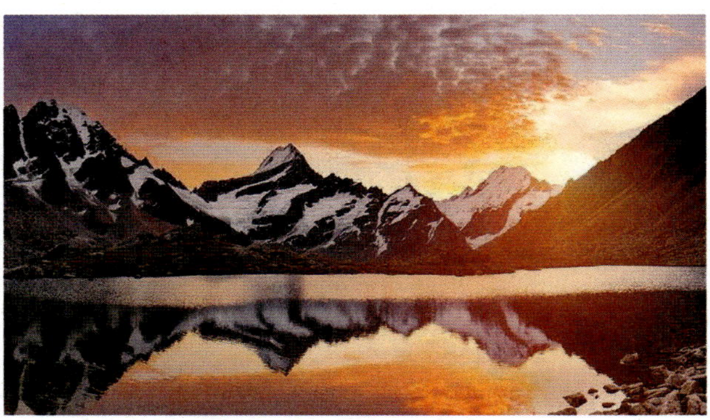

Figure 5.7 'Warm Sunrise Lighting Over a Serene Mountain Landscape' prompt, Adobe Firefly, 2024.

to customize their creations with unique filters and effects. This flexibility makes DeepAI suitable for creators looking to experiment with different visual treatments and build custom aesthetic styles (Gopalan and Kalpinagarajarao, 2023).

Jasper Art (https://www.jasper.ai/image-suite), designed specifically for professional applications, excels at producing original, royalty-free images based on simple text prompts. It supports a wide range of styles, from pixel art and vector designs to 3D renders, making it particularly versatile for projects requiring multiple formats and high customization (Akhmetova and Park, 2023). Artbreeder (https://www.artbreeder.com/), on the other hand, is favored by digital artists for its collaborative nature. Leveraging machine learning models like StyleGAN and BigGAN, Artbreeder allows users to blend and adjust images, creating distinct visuals through parameter customization. This platform's blend-and-merge capabilities enable artists to experiment with variations, which is valuable for exploratory projects where multiple versions are essential (Bertelsen and Short, 2023).

In July 2024, Canva's acquisition of Leonardo.ai demonstrated the increasing significance of generative AI in mainstream design tools. Leonardo.ai enables users to produce high-quality visual assets, including art, images, and videos, with notable speed and style consistency. The platform's accessibility was further enhanced by the release of an iOS app, allowing creators to develop visually stunning content directly on their mobile devices, expanding accessibility and usability for on-the-go projects (Putjorn and Putjorn, 2023). These additional tools emphasize the range and adaptability of generative AI within creative industries. While each model operates on the foundational principles of text-to-image generation, they offer specialized features that make them uniquely suited to different creative needs. Whether for professional, exploratory, or community-driven projects, this growing array of generative platforms provides creators with extensive options to develop and diversify their visual storytelling.

As AI-generated content continues to reshape creative industries, navigating copyright and intellectual property rights has become a crucial step for artists seeking to protect and control their work. Navigating copyright, fair use, and intellectual property is an essential consideration for any creator working with generative AI. As the legal landscape surrounding AI-generated content continues to evolve, many artists are proactively taking steps to retain ownership and control over their creations. One popular approach is to begin with original imagery—such as personal photography, drawings, or paintings—as source material for AI video models. By using proprietary content as input, artists establish a clear foundation of authorship, reducing potential conflicts over copyright and ownership in the final product (Watiktinnakorn et al., 2023).

An increasingly popular method among artists is to create custom training models in tools like Stable Diffusion, using their own artwork as the dataset (Abdikhakimov, 2023). For instance, artist Morgan Harper-Nichols trained a model exclusively on her paintings, ensuring that the generated outputs consistently reflect her distinctive style and creative vision (Hutson and Harper-Nichols, 2023). This approach allows artists to maintain creative control throughout the process, making the generated content both a genuine extension of their existing work and legally more straightforward to claim as original. By training models on proprietary imagery, artists can avoid issues associated with using publicly available datasets, which might contain copyrighted works, and instead produce outputs that are rooted in their own intellectual property.

In addition to maintaining control over the artistic direction, documenting the creative process is essential. By recording each step—from prompt design to image generation and post-processing—creators build a transparent workflow that demonstrates their role in shaping the final product. This detailed documentation can strengthen claims of authorship, making it more likely that the AI-generated work qualifies for copyright protection. Courts and copyright offices are increasingly interested in establishing the level of human involvement in AI-produced works, so such records are a critical asset in any potential legal claims (Abbott and Rothman, 2023).

By combining these strategies—using original images, custom training datasets, and thorough documentation—creators position themselves to assert stronger claims of ownership and copyrightability over their work. This approach not only protects the creator's rights but also supports an ethical framework in which artists can confidently explore generative AI while safeguarding the integrity and authenticity of their creations. As generative technologies continue to expand, these practices offer a pathway for artists to engage with AI responsibly, ensuring that their output remains both legally defensible and aligned with their artistic identity (Atilla, 2024).

5.2 Image-to-Video Generation

Transitioning from static images to animated sequences involves careful planning, especially when working with AI-generated visuals from models like DALL-E 3, Midjourney, Stable Diffusion, and Adobe Firefly. Each of these platforms offers distinctive styles and capabilities, and by aligning image generation with video goals—considering lighting, angles, and consistency—creators can establish a stronger foundation for smooth and cohesive animation. Images created with this forethought will better integrate into video production, reducing the need for extensive adjustments. However, if images lack uniformity in color tones, textures, or stylistic continuity, post-processing in digital imaging software like Photoshop may be necessary. In Photoshop, artists can harmonize various elements, adjust color balances, and add layers or effects that ensure the images align with the visual narrative. When starting with well-prepared images and applying targeted adjustments, creators set up a smoother process for bringing still visuals to life, crafting video sequences that feel both fluid and visually consistent (Farinacci, 2024).

With the foundational images ready, selecting the right type of video generator—whether image-to-video, text-to-video, or video-to-video—becomes essential to achieve the desired creative outcome. In the creative industry, AI-powered video generators serve distinct purposes based on their input modalities. Image-to-video generators animate static images, enabling photographers and visual artists to bring their still artworks to life, adding dynamic elements that enhance storytelling. For instance, platforms like Pika Labs allow users to animate images from sources such as Midjourney, offering features like AI-generated sound effects and upscaling to enrich the visual experience (Ni et al., 2023). Text-to-video generators convert written content into visual narratives, streamlining the production of explainer videos, marketing materials, and educational content. Tools like Synthesia enable the creation of videos with AI avatars by simply inputting text, making it a popular choice for professional training and explainer videos. Video-to-video generators modify existing footage by applying new styles or elements, offering filmmakers and video editors innovative ways to reimagine their content. For example, Runway Gen-3 allows users to transform existing videos by applying new styles or elements, effectively reimagining the original footage. Each type of generator caters to specific creative needs, providing tailored solutions that enhance and streamline the content creation process (Wang et al., 2024). However, it should be noted that many of the examples that follow can be classified in multiple of the three categories.

5.2.1 Pika Labs

Pika Labs (https://pikartai.com/labs/) is an AI-powered platform that enables users to create animated videos from text or image prompts. Developed by Pika Labs, a company focused on advancing the generative capabilities of these new models in creative fields, the platform is designed to democratize video creation, making it accessible to individuals without extensive technical expertise. The platform allows users to input descriptive prompts, which the model, leveraging advanced ML algorithms, interprets to generate dynamic video content. The system operates primarily through its web interface (Fig. 5.8), where users can input text descriptions or upload images to guide the video generation process. Pika Labs utilizes sophisticated AI models to analyze these inputs and produce short video clips that align with the provided prompts. The generated videos typically last about four seconds and run at 24 frames per second, offering a smooth and visually appealing output. The process enables creators to transform static images or textual ideas into animated sequences, enhancing storytelling and visual communication (Zhang et al., 2024).

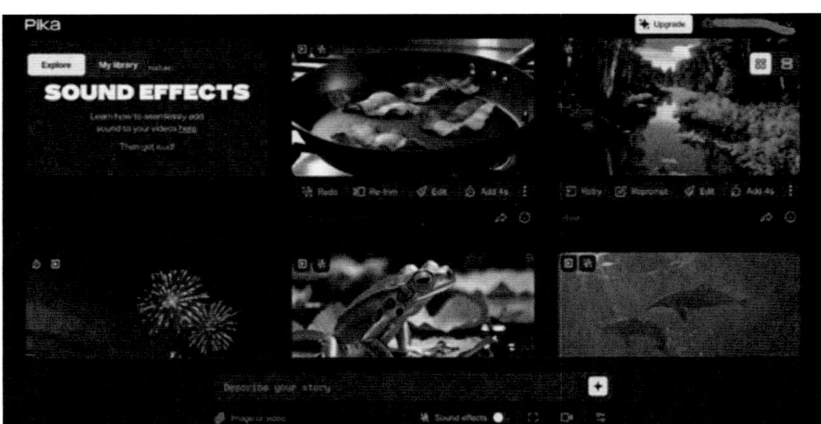

Figure 5.8 Pika Labs AI Interface, Pika AI, 2024.

Common use cases for Pika Labs include creating engaging social media content, developing marketing materials, and producing educational videos. The ability of the platform to quickly generate animated visuals from simple prompts makes it a valuable tool for marketers aiming to capture audience attention, educators seeking to illustrate concepts dynamically, and content creators looking to add motion to their projects. Regarding subscription costs, Pika Labs offers a range of plans to accommodate different user needs. The Basic plan is free and provides 150 monthly video credits, while the Standard plan costs $8 per month (billed yearly) and includes 700 monthly video credits, access to Pika 1.5 and 1.0, fast generations, and the ability to download videos without a watermark for commercial use. Higher-tier plans, such as the Pro and Unlimited options, offer additional credits and features at increased price points. Next, a guide to using the platform shall be presented and summarized in Table 5.5 and can be found on the official site (https://pikartai.com/how-to-use-pika/).

Pika Labs AI is a sophisticated tool designed to democratize animated video creation through text-based and image-based prompts, offering expansive possibilities for visual storytelling and creative expression. Upon signing up via the Pika Labs platform and accessing the main dashboard, users can either select from pre-designed templates or begin a new project. There are extensive tutorials made available by the company (https://pikalabsai.org/pika-art-1-5/). Using text-based prompts, creators input detailed descriptions—such as "sunlit forest path at dawn with mist and soft lighting"—which the system interprets to generate a smooth, four-second animated

Table 5.5 Guide to generating videos using Pika Labs AI.

Step	Description	Details
1. Sign Up and Log In	Create an account on Pika Labs.	Visit the Pika Labs website and sign up using your Google or Discord account.
2. Access the Platform	Navigate to the main interface.	After logging in, you'll be directed to the dashboard where you can start creating videos.
3. Choose a Template or Start from Scratch	Begin your project.	Browse available templates or select the option to create a new project.
4. Input Your Prompt	Provide text or image prompts.	Enter a descriptive text prompt or upload an image to guide the AI in generating your video.
5. Customize Video Settings	Adjust parameters.	Use the settings to modify aspect ratio, frames per second (fps), and motion controls to achieve the desired effect.
6. Generate the Video	Create the video.	Click the 'Generate' button and wait for the AI to process your input and produce the video.
7. Review and Edit	Assess the output.	Watch the generated video and make any necessary adjustments by refining your prompt or settings.
8. Export the Final Video	Save your creation.	Once satisfied, export the video in your preferred resolution and format.

clip at 24 frames per second. Pika Labs also supports customization of video settings, including adjustments to aspect ratio, frame rate, and motion dynamics, ensuring that users can align the visual output with their specific creative objectives. This flexibility makes Pika Labs a valuable tool for both novice and experienced users seeking high-quality, animated visual content.

A structured approach is essential to maximizing the output quality of videos generated with Pika Labs. Detailed prompt construction, which specifies key visual elements—such as color, lighting, and environment—can yield highly tailored animations. For instance, a prompt like "high-speed urban street at night with neon reflections and a light rain effect" results in a more layered visual than a basic descriptor. Utilizing Pika Labs' negative prompting feature to exclude unwanted elements further refines the output, particularly for complex scenes. Motion and frame rate adjustments offer additional control: tranquil settings benefit from reduced motion for a contemplative feel, while action-intensive visuals may achieve enhanced dynamism through higher motion values. An iterative approach, involving multiple prompt refinements and preview assessments, is recommended to progressively refine outputs and achieve the desired effect.

In addition to text-to-video capabilities, Pika Labs offers an image-to-video feature, allowing users to animate static images by uploading a picture and applying descriptive prompts to guide animation. For example, an uploaded image of a serene lake can be animated with a prompt like "gentle ripples and soft morning light," bringing dynamic motion to a previously static scene. Experimentation with prompt variations and visual settings is encouraged to optimize the aesthetic and narrative quality of the animated video. Engaging with the Pika Labs user community on platforms like Discord also offers valuable insights into advanced prompt strategies and troubleshooting techniques, enabling users to refine their skills and expand their creative range. The integration of detailed prompts, systematic experimentation, and community collaboration allows users to leverage Pika to produce visually compelling, affordable, and adaptable videos suited to various professional and artistic applications.

5.2.2 Kaiber

Kaiber (https://kaiber.ai/), established in 2023, represents an innovative entity within the creative technology sector, dedicated to enhancing the collaborative potential between human creativity and artificial intelligence. As a company, Kaiber operates with a mission to support global artists and creators, equipping them with advanced tools that transform conceptual ideas into polished visual content. The Kaiber team comprises specialists in art, research, and technology, unified in their commitment to leveraging the capabilities of the latest generation of intelligent systems for artistic innovation. The core offering of the company, Superstudio (Fig. 5.9) (https://www.kaiber.ai/superstudio/), exemplifies this mission by serving as an AI-native platform that integrates cutting-edge generative tools, thus facilitating new modes of interaction between creators and machine creativity (Cho et al., 2024).

Superstudio distinguishes itself by providing a seamless, user-friendly interface that amalgamates leading AI image and video models, including Luma Lab's

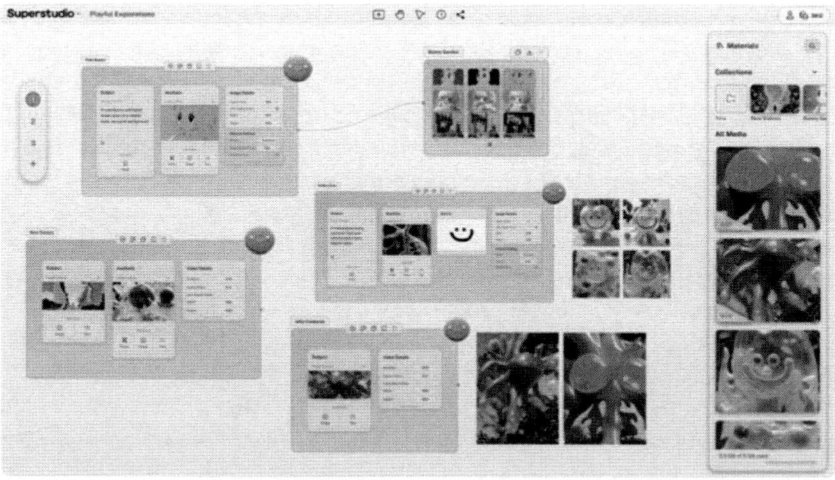

Figure 5.9 Superstudio Interface, Kaiber, 2024.

Dream Machine (https://lumalabs.ai/dream-machine) and Black Forest Labs' Flux (https://blackforestlabs.ai/). Within this platform, users engage with the model through a structured canvas, blending their creative vision with AI outputs to produce sophisticated visual narratives. Functionally, Superstudio supports a diverse array of content generation types: it encompasses text-to-video generation, which translates textual prompts into dynamic video sequences; image-to-video animation, allowing for the transformation of static images into animated scenes; and audioreactivity, wherein visual elements respond to synchronized audio inputs such as music tracks or voiceovers. Furthermore, Superstudio provides customizable animation controls, enabling users to dictate camera movements, scene transitions, and other cinematic elements, making it an effective tool for users seeking refined, AI-assisted storytelling.

Kaiber has gained significant recognition for its application across multiple use cases, particularly in industries requiring high-quality, visually compelling outputs. Among its most common applications, Kaiber is frequently employed to produce music videos with dynamic visual elements that respond directly to audio cues, creating immersive audiovisual experiences. In marketing, the platform functionalities allow for the creation of visually distinctive advertising content that captures audience attention and engages viewers. Additionally, the tool supports educational content creators by providing tools for animated explanatory visuals and is popular among social media creators who seek unique, stylistically engaging videos to enhance audience engagement. With video upscaling options that extend to 1080 p and 4 K resolutions, Kaiber ensures professional-grade output quality, enabling users to push creative boundaries while maintaining rigorous production standards (Lyu et al., 2024).

The platform offers a structured subscription model that supports diverse user needs, with varying levels of access and functionality. The introductory Explorer Plan, priced at $5 per month, includes 300 monthly credits and access to

Flipbook (for up to one-minute videos), Motion (up to 16 seconds), and Transform (up to one minute). The Pro Plan, at $10 per month, expands these capabilities with 1,000 monthly credits, enabling video lengths of up to four minutes and video upscaling options to 1080 p and 4 K resolutions. For more advanced usage, the Artist Plan at $25 per month provides 2,500 credits, along with all Pro Plan features and beta access to new platform functionalities. Users may also choose annual subscriptions to save up to 33% compared to monthly rates. A free plan, including 200 credits, allows users to explore basic features without cost. Next, a guide to using the platform shall be presented and summarized in Table 5.6 and can be found on the official site (https://www.kaiber.ai/product/).

Creating videos with Kaiber's Superstudio involves a streamlined, user-centric process that blends advanced AI functionality with intuitive design, making it suitable for diverse creators, from artists to marketers. The journey begins with account creation and a tour of Superstudio's workspace, known as the Canvas, which serves as the interactive hub where users can design, test, and refine their visual outputs. Superstudio supports both text-to-video and image-to-video generation modes,

Table 5.6 Guide to generating videos using Kaiber.

Step	**Action**	**Details**
1. Create an Account	Sign up on Kaiber's website.	Visit Kaiber's Superstudio and register using your email.
2. Access Superstudio	Log in to your account.	After registration, log in to access the Superstudio interface. Kaiber Help Center
3. Navigate the Canvas	Familiarize yourself with the workspace.	Explore the Canvas, which serves as your creative playground for generating and editing media.
4. Generate Text-to-Video	Use the Luma Video Flow.	- Click on the **Luma Video Flow** in the navigation bar to add it to the Canvas. - Enter a descriptive text prompt to define your video's content. - Optionally, add Start and End Keyframe images to guide the video's progression. - Click 'Generate' to create a 5-second video.
5. Generate Image-to-Video	Utilize the Video Lab Flow.	- Click on the **Video Lab Flow** in the navigation bar. - Upload an image to serve as the video's starting point. - Enter a text prompt to describe the desired animation or transformation. - Click 'Generate' to produce the video.
6. Review and Edit	Assess the generated media.	- Play the generated video to ensure it meets your expectations. - If adjustments are needed, modify the prompts or settings and regenerate.
7. Export the Final Output	Save your creation.	- Once satisfied, download the video in your preferred format and resolution.

offering flexibility in content creation. Text-to-video functionality allows users to enter descriptive prompts that guide the generation of a short video, which can be further enhanced by adding Start and End Keyframe images to influence transitions. Similarly, image-to-video generation begins with an uploaded image, allowing the AI to animate or transform it based on an accompanying descriptive prompt. This functionality allows creators to develop narrative coherence in visual storytelling while experimenting with both static and dynamic input types, making it versatile for applications such as promotional videos, educational content, and multimedia art.

For optimal results, Superstudio users should carefully consider how they construct and refine their prompts, as Kaiber's AI responds best to well-defined, descriptive input. For instance, a prompt like "sunset beach with waves gently rolling" yields more targeted visuals than vague phrases, ensuring that the model generates a coherent scene. Additionally, users may integrate Start and End Keyframes for a defined progression, enhancing the video's narrative structure or visual flow. Adjusting camera angles, frame rates, and scene transitions within the Canvas allows for greater control over the pacing and mood of the video, contributing to an immersive viewer experience. While the platform is designed to interpret prompts effectively, users benefit from iterative testing—making slight adjustments to prompts and settings to refine output progressively. This practice not only ensures that the final result aligns with the vision of the creator but also allows exploration of the range of the tool, from naturalistic visuals to surreal or abstract aesthetics.

Superstudio offers a suite of best practices for users aiming to maximize the creative potential of the platform. One essential strategy is balancing experimentation with the model supporting Superstudio and following prompt structuring guidelines available in Kaiber's resources (https://blog.kaiber.ai/blog/how-to-prompt-in-kaiber-for-beginners); this helps in avoiding common pitfalls, such as ambiguous visuals or unintended image overlays. Using descriptive prompts with carefully chosen modifiers—like "dawn light, misty forest, serene"—enhances output clarity and guides the AI toward desired stylistic choices. Additionally, credit management is key, as each video generation consumes a specific amount; users are encouraged to plan their generative explorations thoughtfully, especially when working with detailed, high-resolution outputs. Engaging with Kaiber's user community (https://www.instagram.com/kaiber.ai/), including online forums and tutorials, can also provide valuable insights, offering examples of successful prompt structures and troubleshooting methods. By adhering to these practices, creators can leverage Superstudio to produce polished, visually compelling videos that capture their artistic intent or brand message with precision.

5.2.3 Runway Gen-3

Runway Gen-3 Alpha (https://runwayml.com/research/introducing-gen-3-alpha) (Fig. 5.10) stands at the forefront of AI-driven video generation, representing a marked advancement in fidelity, consistency, and motion refinement. Developed by Runway, an innovative company specializing in applied AI research, Gen-3 Alpha builds upon the foundation of its predecessors to offer enhanced functionality for diverse creative needs. This model accepts multiple input modalities, including both

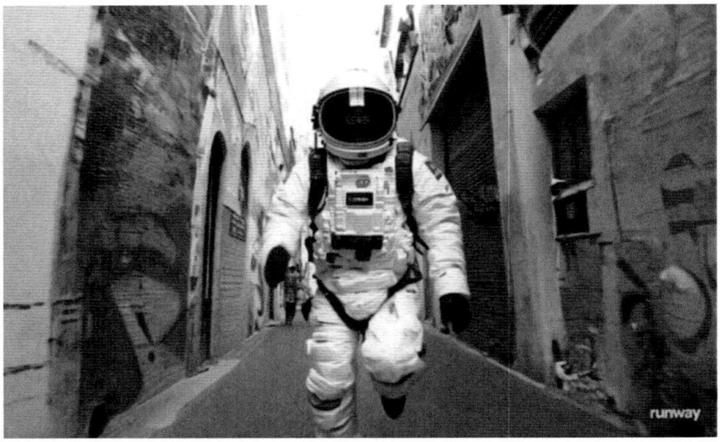

Figure 5.10 Astronaut Output Example, Runway Gen-3 Alpha, 2024.

text and image prompts, allowing users to seamlessly transition from conceptual ideas to sophisticated video outputs. The interpretation of these varied input types allows Gen-3 Alpha to generate video content that adheres more closely to the vision of a creative, making it a highly versatile tool for professional filmmakers, marketers, and visual storytellers seeking to integrate AI-generated content into their workflows (Ma, 2024).

Runway, the organization behind Gen-3 Alpha, is dedicated to expanding creative horizons through artificial intelligence, with a core focus on developing generative tools for professionals in media, advertising, visual effects, and digital content creation. As one of the most popular platforms in industry for integrating generative visuals into video production, Runway has consistently pioneered tools that redefine traditional production processes. In addition to Gen-3 Alpha, the company has released an accessible iOS app, expanding its reach to a broader consumer base. The Gen-3 model exemplifies Runway's commitment to innovation, as it supports both text-to-video and image-to-video functionalities, and further enhances the video fidelity and motion quality established by Gen-2. These capabilities are invaluable for a variety of applications, from film production and advertising to educational content and brand storytelling, where seamless, AI-enhanced visual narratives are increasingly in demand.

Despite the tool's impressive capabilities, the cost associated with high-quality AI video production on Runway remains a significant consideration. Runway's subscription model is tiered to accommodate varying levels of usage: the Basic Plan is free, offering 125 credits per month with limitations to 720 p resolution and watermarked outputs, suitable for small projects like four seconds of Gen-1 (video-to-video) or 16 seconds of Gen-2 (text-to-video). The Standard Plan, priced at $12 per user per month, includes 625 credits, supporting up to 125 seconds of Gen-2 video and 4 K exports, with 100 GB storage and no watermarks. For more advanced needs, the **Pro Plan** at $28 per month offers 2,250 credits and additional features such as custom AI model training and 500 GB of storage, sufficient for around 450 seconds of Gen-2 video. The **Unlimited Plan** provides unrestricted video generation at $76 per user per month, ideal for high-output projects. While Runway's flexible pricing model allows for adaptability, the production of longer, high-resolution films remains cost-prohibitive for independent creators. Upgrading to higher plans or purchasing additional credits may be necessary for large-scale projects, underscoring the ongoing challenge of balancing quality and budget in AI-powered video production. Next, a guide to using the platform shall be presented and summarized in Table 5.7 and can be found on the official site at the Runway Academy (https://academy.runwayml.com/gen3-alpha/using-image-to-video-in-gen3-alpha).

Using Runway Gen-3 Alpha, an advanced generative AI tool, involves a structured process that enables users to produce high-quality, AI-driven video content. Beginning with the creation of an account on Runway's platform, users are introduced to a versatile interface where they can experiment with text-based, image-based, or hybrid video generation methods. Once logged in, users navigate the "Text/Image to Video" function to initiate a project, selecting either a text-based prompt or uploading an image as a reference (https://youtu.be/JkPO-mBV9bQ). For optimal results, the text prompts should be carefully constructed to convey specific visual elements, such as environment, subject details, and movement, thereby guiding the model to produce content that aligns with a specific vision. Optional image uploads provide additional guidance for the platform, ensuring fidelity to a desired aesthetic or style. This latest model also has a more robust camera control (https://www.youtube.com/watch?v=0buDtZKLDJ8). Configuring

Table 5.7 Guide to generating videos using Runway Gen-3 Alpha.

Step	Action	Details
1. Create an Account	Sign up on Runway's website.	Visit Runway's website and register using your email.
2. Access Gen-3 Alpha	Log in to your account.	After registration, log in to access the Gen-3 Alpha interface.
3. Navigate to Text/ Image to Video	Select the appropriate tool.	From your Runway dashboard, click on "Text/Image to Video" to start generating videos. Runway Academy
4. Input Text Prompt	Describe your desired video.	Enter a detailed text prompt that conveys the scene, subject, and camera movements you envision.
5. Upload an Image (Optional)	Provide a reference image.	If desired, upload an image to guide the video generation process.
6. Configure Settings	Set video parameters.	Choose the video length (5 or 10 seconds) and resolution (typically 1280 × 768).
7. Generate Video	Initiate the creation process.	Click the "Generate" button to start the video generation.
8. Review and Edit	Assess the generated video.	Once the video is generated, review it and make any necessary adjustments to the prompt or settings.
9. Export the Final Output	Save your creation.	After finalizing, download the video in your preferred format and resolution.

video length, generally between 5 and 10 seconds, and resolution further refines the output, allowing for customization aligned with professional standards (https://youtu.be/B_RfJ1B5wME).

The creation process within Gen-3 Alpha is driven by a series of steps that prioritize iterative refinement. Once a prompt and any optional images have been inputted, users generate the video by selecting the "Generate" option, which initiates the AI's interpretative processes. Reviewing the generated video is a critical step, as it allows users to assess alignment with their vision and adjust settings or prompts as needed. This iterative approach, where creators refine prompts or modify parameters after each generation, ensures that each video iteration progressively aligns closer to the intended outcome. Upon satisfactory completion, users can download the final product in the desired resolution, often utilizing 4 K exports for high-quality, professional use. This process of refinement and iteration exemplifies Gen-3 Alpha's flexibility in allowing users to experiment with both structured and spontaneous visual storytelling, enabling diverse applications across industries.

For best results, users should adhere to several recommended practices. Clarity in prompt construction is essential; specific descriptors—such as "desert landscape at sunset with drifting sand and soft shadows"—help guide the AI in generating content that closely matches the creator's vision. Additionally, mindful management of credit usage is critical, particularly for large or complex projects, as each video generation session consumes credits from the user's subscription. Users benefit from a well-planned approach that minimizes unnecessary iterations and conserves credits. To optimize results, experimentation with diverse prompts and configurations is encouraged; Runway's flexible parameters allow for exploration within the creative process, with each variation offering new potential insights into the capabilities of the platform. Engaging with the available resources, such as the platform's guidance on prompt structuring and AI interpretation, further enhances the user experience. Through strategic experimentation, credit management, and structured prompt refinement, creators can maximize Gen-3 Alpha's capabilities to produce refined, compelling visual narratives tailored to professional needs (Naji, 2024).

In sum, the exploration of Pika Labs, Kaiber, and Runway Gen-3 Alpha highlights the evolving sophistication and range of AI-driven video generation platforms available for diverse creative applications. Each platform offers unique strengths: Pika Labs democratizes video animation with accessible tools for image and text-based prompts, making it a valuable asset for social media content, marketing, and education; Kaiber combines intuitive design with dynamic audio-reactive and customizable animation features, supporting visually rich content for artistic and commercial use; and Runway Gen-3 Alpha stands out as an industry-leading tool for high-fidelity, flexible video generation, accommodating text, image, and video inputs to meet professional demands. While each platform provides unique functionalities, users across all three benefit from thoughtful prompt construction, iterative refinement, and effective resource management to achieve optimal results. As the tools and techniques in image-to-video generation continue to advance, the next section delves into text-to-video generation, examining how AI can transform textual input into complex visual narratives, further expanding the creative possibilities within generative media.

5.3 Text-to-Video Generation

This section addresses the advancements and applications of text-to-video generation within the realm of AI-powered media tools. Unlike image-to-video generation, which often starts from a static visual or guided image prompt, text-to-video generation enables creators to produce video content directly from textual descriptions (Liao et al., 2024). This approach opens new avenues for content creation, particularly in areas such as education, corporate training, and digital marketing, where quick, scalable video production is in demand. Text-to-video platforms interpret text prompts and convert them into videos featuring animated elements, simulated environments, or even virtual avatars that narrate or present content (Wang et al., 2024). The three tools discussed in this section—Synthesia, Pictory, and Elai.io—are leading platforms in this domain, each designed to facilitate the generation of high-quality, AI-enhanced video content tailored to different user needs and contexts.

Similar to image-to-video tools, these text-to-video platforms often support multiple types of input, allowing users to blend text with images, audio, or even pre-existing video clips to enhance the final output. For instance, while Synthesia is known for its advanced use of digital avatars, enabling realistic narration for training or instructional videos, Pictory emphasizes ease of use for creating video summaries or content highlights from long text sources, making it ideal for marketers and content creators. Elai.io, on the other hand, offers customizable virtual presenters and interactive features, providing an interactive and personalized touch to video content. Each platform brings unique strengths to text-based video production, catering to a range of creative and professional applications where rapid, adaptive video generation is essential. As this section unfolds, it will examine the capabilities, primary use cases, and best practices associated with each of these tools, highlighting the potential of text-to-video AI to redefine digital content creation.

5.3.1 Synthesia

Synthesia (https://www.synthesia.io/) is an advanced AI-powered platform that transforms text into professional-quality videos, catering to industries that require scalable, adaptable content. Established in 2017 by Victor Riparbelli and Steffen Tjerrild, Synthesia was developed to make high-quality video production accessible and cost-effective, reducing reliance on traditional filming equipment, actors, and costly production setups. The platform uses proprietary algorithms to create lifelike digital avatars capable of narrating scripts in over 140 languages, a feature particularly advantageous for global enterprises and educational institutions (Muñoz-Basols et al., 2023). The technology behind the system is known for producing realistic avatars with accurate lip-syncing, allowing for seamless and polished visual narratives from simple text inputs.

The versatility of the platform has led to its widespread application across various domains, that were already using video for various purposes, including corporate training, marketing, and education (Díaz Redondo et al., 2021; Noetel et al., 2021). In corporate environments, Synthesia enables the creation of onboarding materials, training modules, and internal communications that are easily scalable and accessible across international teams (Lind, 2024). For marketers, the system allows for the production of personalized video messages, product demonstrations, and promotional content tailored to specific audiences, thus enhancing engagement without extensive resource allocation (Martínez-Rolán et al., 2023). Educational institutions and e-learning platforms use it to generate interactive, instructional videos that can facilitate learning in diverse linguistic contexts, improving accessibility for a global student body (Joseph, 2023).

The capabilities of the platform center around its core text-to-video generation functionality, allowing users to input a script, select an avatar from an extensive library, and customize the video through various templates (https://www.synthesia.io/video-templates) and backgrounds. Additionally, Synthesia supports video localization, where content is translated and lip-synced accurately to suit multilingual audiences—a feature highly valuable for multinational corporations and global outreach programs (Flew et al., 2024). The tool also provides custom avatar creation, enabling users to create a unique digital representative of their brand or organization. With features like intelligent voice synthesis and adaptable video templates, Synthesia allows users to produce high-impact, branded video content tailored to their communication and presentation needs. This functionality enhances the ability of organizations to convey messages with consistency and personalization, leveraging the precision of machine creativity in the platform to maintain brand standards.

Finally, pricing for the model reflects its flexibility in serving diverse client needs, ranging from small projects to large-scale corporate operations. The platform offers a Free Plan with limited access—allowing 3 minutes of video per month and six stock avatars—to enable initial exploration of its features. For more extensive needs, the Starter Plan at $29 per month includes 10 video minutes, access to a broader range of avatars, and greater flexibility in video customization. The Creator Plan, priced at $89 per month, expands this capacity to 30 video minutes, offering advanced customization and additional support, which is suitable for small to medium-sized projects. For enterprises, there is an Enterprise Plan with custom pricing and unlimited video minutes, designed to support large organizations with extensive video production needs. This tiered pricing structure allows users to align their subscription level with their content demands, making the system an accessible solution for businesses of varying scales while maintaining a focus on quality and customization capabilities. Next, a guide to using the platform shall be presented and summarized in Table 5.8 and can be found on the official site at the Synthesia Academy (https://www.synthesia.io/academy).

Employing Synthesia for AI-driven video creation follows a systematic yet flexible process, allowing users to transform text into high-quality, tailored video content efficiently. This process begins with establishing an account on Synthesia's website, providing access to various video templates and customization features designed for professional and creative projects alike. After logging in, creators initiate a new video project by selecting "Create Video" on the dashboard, which

Table 5.8 Guide to generating videos using Synthesia.

Step	Action	Details
1. Create an Account	Sign up on Synthesia's website.	Visit Synthesia's website and register using your email.
2. Access the Platform	Log in to your account.	After registration, log in to access the Synthesia interface.
3. Start a New Project	Click on "Create Video."	From your dashboard, select the "Create Video" option to begin a new project.
4. Choose a Template	Select a video template.	Browse and choose a template that fits your video's purpose.
5. Input Script	Enter your text content.	Type or paste your script into the text box provided.
6. Select an AI Avatar	Choose a digital presenter.	Pick an AI avatar to narrate your script.
7. Customize Video	Adjust settings and visuals.	Modify background, text appearance, and other elements to suit your brand.
8. Preview Video	Review the generated video.	Click "Preview" to watch the video and ensure it meets your expectations.
9. Edit if Necessary	Make adjustments.	If needed, edit the script, avatar, or visuals based on the preview.
10. Generate and Export	Finalize and download.	Once satisfied, click "Generate Video" and, after processing, download the final video.

presents them with an assortment of templates suited for specific applications, such as instructional modules, promotional content, or corporate communications. The tool allows users to input their script directly into a text field, enabling the AI to generate a narrative delivered by a chosen digital avatar (Fig. 5.11). Selecting an appropriate avatar is critical for aligning the video's presentation style with its intended audience, ensuring a personalized and effective delivery of the content.

Configuring visual elements within the interface is essential for aligning the aesthetic and message of the video with the organizational branding or unique project requirements. Creators have the option to modify backgrounds, text formatting, and overall layout to ensure consistency with brand identity and enhance viewer engagement. Previewing the video during these adjustments serves as a quality checkpoint, allowing users to evaluate the clarity of the narration, avatar presentation, and visual structure. Synthesia supports an iterative refinement process, which lets users make continuous adjustments to the script, visual elements, and other settings until the video meets their standards. Upon completing these revisions, creators can finalize the video, allowing Synthesia to render the final product for download. The iterative nature of the platform underscores the importance of previewing and refining to achieve a seamless and polished output.

To optimize video creation with Synthesia, several best practices are recommended, including constructing clear and direct scripts, choosing avatars

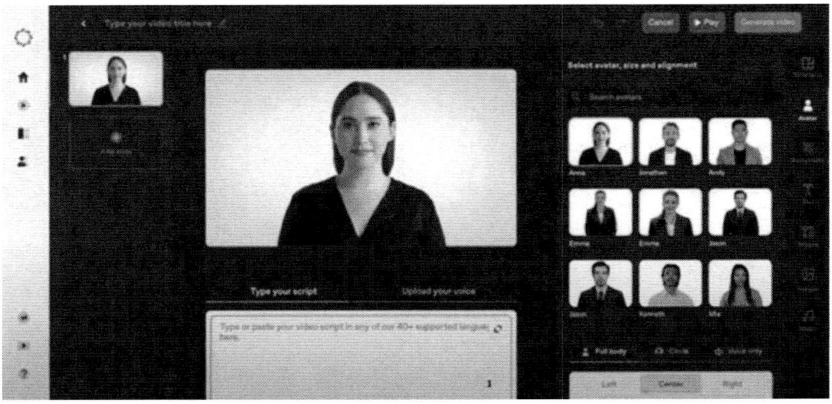

Figure 5.11 Synthesia Dashboard for Creating AI Avatars, Synthesia, 2024.

thoughtfully, and making strategic use of customization options. For instance, crafting a concise, precise script enhances the delivery, ensuring that key messages are conveyed effectively and professionally. Moreover, selecting an avatar that resonates with the intended audience can boost engagement by adding relatability and a personal touch to the video. The multilingual support for the platform, encompassing over 140 languages, further broadens its applicability, making Synthesia particularly beneficial for multinational organizations and educational programs aiming to enhance accessibility. Users should also consider the various subscription levels, as each video session utilizes specific credits, which can be limited depending on the plan. Selecting a subscription plan that aligns with the project's scope is advisable, especially for larger or ongoing content needs. Thus, combining clear scripting, customization, and mindful resource management allows creators to maximize the capabilities of Synthesia, producing impactful, professional videos that serve diverse communication and branding objectives.

5.3.2 Pictory

Pictory (https://www.pictory.ai) represents a transformative tool in AI-driven video creation, designed to simplify and expedite the process of converting long-form text-based content into visually engaging video formats. Created by a team of experienced entrepreneurs and technologists, this platform aims to make high-quality video production accessible to users without requiring specialized technical skills or extensive resources. Through its use of artificial intelligence, Pictory automates core aspects of the video creation process, including editing, captioning, and adding appropriate visuals. This automation reduces the time and effort traditionally required for producing professional video content, allowing individuals and businesses to repurpose blogs, articles, and other long-form content into shareable videos that resonate with modern audiences (Kaur et al., 2024). The ease of use of the platform and emphasis on efficiency make it a powerful resource for those seeking to enhance their content strategies with minimal manual intervention.

The application scope of this platform is broad, extending its utility across various domains such as marketing, education, and corporate communication. Content creators and marketers frequently employ this AI-powered tool to convert written material into video snippets optimized for social media platforms, helping increase viewer engagement and reach. Educators and trainers use the platform's capabilities to craft instructional videos, transforming textual information into interactive, visually dynamic learning resources that can improve knowledge retention and accessibility (Hadi and Ainy, 2024). Additionally, businesses benefit from using the software for creating promotional videos, product demonstrations, and internal communications, leveraging its AI-driven features to maintain a professional video aesthetic efficiently (Isler et al., 2023). With functionalities such as automatic transcription, subtitle addition, and integration of stock footage and audio, this video creator supports users in crafting videos that are visually polished and thematically aligned with their content objectives.

The platform offers flexible pricing models to accommodate a range of user needs, beginning with a 14-day Free Trial that allows new users to explore its core features and create up to three video projects. The Starter Plan, priced at $19 per month, includes 30 video projects per month, with each video up to 10 minutes in duration, access to a library of over 2 million royalty-free videos, and 34 text-to-speech AI voices. For users with more extensive requirements, the Professional Plan at $39 per month provides up to 60 videos monthly, extends video length to 20 minutes, grants access to over 12 million stock videos, and introduces bulk video download options along with multiple brand kits for consistent branding across projects. Finally, the Teams Plan at $99 per month is designed for collaborative workflows, offering up to 90 videos monthly, each up to 30 minutes, along with team collaboration tools and API access for organizational integration. These pricing tiers allow users to select the plan that best aligns with their video production needs and organizational scale, making this tool adaptable for both individuals and teams engaged in regular, high-volume content creation. Next, a guide to using the platform shall be presented and summarized in Table 5.9 and can be found on the official site (https://kb.pictory.ai/en/).

Using the AI-driven platform to create video content from text-based or existing media is an accessible yet comprehensive process that allows for efficient video generation. The initial step requires account registration on Pictory's website, after which users can log in to the platform's main dashboard. From the dashboard, users begin a new project by selecting "Create New Project," which opens up options to choose a content source. Pictory provides multiple input methods, including text-based scripts, URLs of existing articles, or previously captured media files, catering to various content creation needs. By selecting an input source, users can adapt their content format to suit the intended purpose of the video, whether for educational, promotional, or informational applications. Entering the text or uploading content initiates the video generation process, where the platform begins transforming these inputs into cohesive video scenes.

Customization is a critical component in Pictory, allowing users to tailor scenes to enhance engagement and align with branding guidelines. Once Pictory generates

Table 5.9 Guide to generating videos using Pictory.

Step	Action	Details
1. Account Registration	Sign up on Pictory's website.	Visit Pictory.ai and register using your email address to create an account.
2. Log In	Access your account.	Use your registered email and password to log in to the Pictory platform.
3. Start a New Project	Initiate video creation.	Click on "Create New Project" to begin the video creation process.
4. Select Input Method	Choose content source.	Decide whether to create a video from a script, article, or existing media.
5. Input Content	Provide text or media.	Enter your script, paste the article URL, or upload media files as per your selection.
6. Customize Scenes	Edit visual elements.	Modify scenes by selecting visuals, adjusting text, and adding voiceovers or background music.
7. Preview Video	Review the project.	Use the preview function to watch the video and ensure it meets your expectations.
8. Make Adjustments	Refine content.	Edit any elements that require changes based on the preview.
9. Generate Video	Finalize creation.	Once satisfied, click "Generate" to produce the final video.
10. Download and Share	Access the output.	After processing, download the video file and share it on your desired platforms.

an initial version, users can edit visual elements by selecting from available stock footage, modifying scene transitions, and adding background music or voiceovers (Fig. 5.12). The platform provides an extensive library of AI-generated voiceover options in addition to an option for users to upload custom voice narrations, offering a versatile approach to voice customization. Users are encouraged to utilize Pictory's brand kit feature, which enables the incorporation of specific fonts, colors, and logos, ensuring consistent branding across all video outputs. A thorough preview phase is available to review the video's pacing, scene transitions, and visual quality. This preview process is vital for identifying any necessary adjustments, such as refining script length to maintain viewer engagement or editing scene duration to improve clarity.

Adhering to best practices in Pictory can optimize the platform's output quality and effectiveness. For example, maintaining scene duration under ten seconds helps keep viewer attention, while well-structured scripts ensure concise and impactful messaging. Users should also be mindful of scene transitions and visual consistency, selecting relevant visuals that enhance comprehension without overwhelming the narrative flow. Leveraging Pictory's extensive voiceover options allows for tailored narration that can resonate with the intended audience, particularly in educational or professional contexts. Resource management, such as using the brand kit to maintain aesthetic uniformity, strengthens brand identity, ensuring a polished final product. The final step, generating and downloading the video, enables easy sharing across

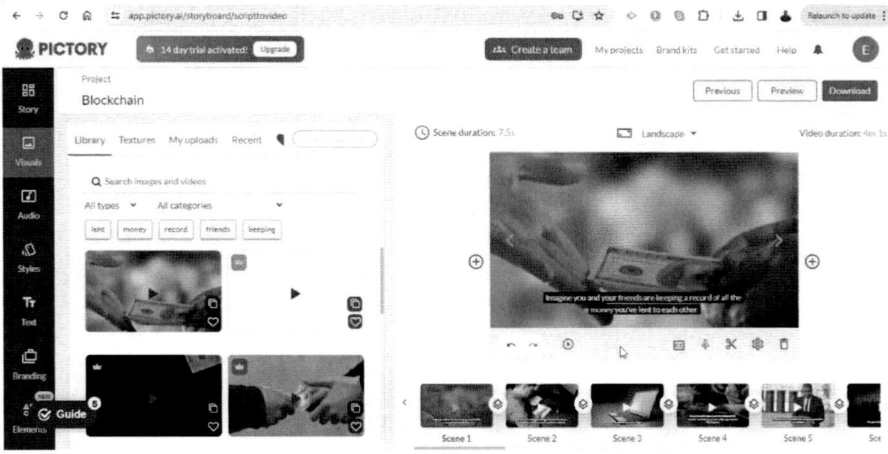

Figure 5.12 Pictory Dashboard, Pictory, 2024.

social media or integration within a broader content strategy. By following these steps and applying best practices, users can effectively create visually engaging, AI-powered video content tailored to diverse audiences and communication objectives.

5.3.3 Elai.io

Elai.io (https://elai.io) is an advanced AI-driven platform that allows users to transform text-based content into high-quality, visually engaging videos. Established in 2021 by a team of technology enthusiasts with deep-tech expertise, Elai.io has its roots in Ukraine and has expanded to include a global team dedicated to pushing the boundaries of AI in video production. The platform uses sophisticated algorithms to create lifelike digital avatars that narrate scripts in over 75 languages, effectively removing the need for traditional recording studios, actors, or specialized equipment. With its accessibility and focus on automation, Elai.io provides users from various sectors the capability to efficiently produce professional-grade videos, making it a valuable tool for both individual creators and large organizations seeking scalable content solutions (Militsyna, 2023).

Elai.io has been widely adopted across several industries, with primary use cases in corporate training, marketing, and education. Within corporate environments, the platform serves as an efficient medium for developing training modules, onboarding videos, and internal communications that can be distributed across international teams. Marketing professionals utilize Elai.io to create customized video messages, product demonstrations, and promotional content tailored to diverse customer bases, enhancing engagement through personalized, AI-driven visuals. In the education sector, e-learning platforms and academic institutions employ Elai.io to generate interactive instructional videos that enrich the learning experience for students. By automating video production, Elai.io enables these sectors to produce content that is not only visually cohesive but also linguistically adaptable, meeting the needs of global audiences with minimal manual intervention.

Elai.io's pricing structure is designed to cater to different levels of user engagement, offering plans that scale with the scope and frequency of video production needs. The Free Plan provides users with a limited introductory experience, including 1 minute of video credit (with one slide per render), access to automated translations, and over 25 avatars. For users with moderate production needs, the Basic Plan is priced at $23 per month, offering 15 video minutes monthly, API access, and additional minutes purchasable at $1.54 per minute. The Advanced Plan, at $100 per month, includes 50 video minutes, premium voices, custom music upload, and support for Ultra HD 4 K video avatars, with additional minutes available at $2.00 per minute. Finally, Elai.io offers a Custom Corporate Plan for large organizations, which includes a tailored setup, all advanced features, a dedicated account manager, and unlimited user access, with pricing customized to specific organizational requirements. This tiered approach allows Elai.io to accommodate a wide range of user profiles, from individual content creators to enterprises requiring large-scale, high-resolution video production capabilities. Next, a guide to using the platform shall be presented and summarized in Table 5.10 and can be found on the official site, Elai.io Lab (https://elai.io/academy/).

Table 5.10 Guide to Generating Videos using Elai.io.

Step	Action	Details
1. Account Registration	Sign up on Elai.io's website.	Visit Elai.io and register using your email address to create an account.
2. Log In	Access your account.	Use your registered email and password to log in to the Elai.io platform.
3. Start a New Project	Initiate video creation.	Click on the "Create Video" button to begin a new project.
4. Choose a Template	Select a video template.	Browse and choose a template that fits your video's purpose, or start from scratch.
5. Input Script	Enter your text content.	Type or paste your script into the text box provided for each slide.
6. Select an AI Avatar	Choose a digital presenter.	Pick an AI avatar from the sidebar to narrate your script; position and size the avatar as needed.
7. Customize Slides	Adjust settings and visuals.	Modify backgrounds, add images, shapes, animations, and adjust text to align with your brand.
8. Preview Video	Review the generated video.	Click the "Play" button next to the text-to-speech box to listen to the speech; full video preview requires rendering.
9. Render Video	Finalize creation.	Once satisfied, click the "Render" button; the video will be processed, and you'll receive a confirmation email upon completion.
10. Download and Share	Access the output.	After rendering, download the video file and share it on your desired platforms.

Using the AI-driven platform for video creation is a structured process that enables users to efficiently produce high-quality, personalized video content from text scripts. The initial step involves registering for an account on the company website, which grants access to a range of video templates and customization options. Once logged in, users can start a new video project by selecting "Create Video" from the dashboard. Elai.io offers a variety of pre-designed templates tailored to common use cases—such as instructional or promotional videos—or the option to start from scratch for a fully customized approach. After selecting a template, users enter their script into the text fields, with each section or slide corresponding to a segment of the video. The platform's AI interprets the script and assigns narration accordingly, allowing users to convey information through a virtual avatar, enhancing the personalization and engagement of the content.

Customization is a central feature within Elai.io, enabling users to align their videos with brand standards or specific stylistic requirements. By selecting an AI avatar from their library, users can choose a digital presenter that resonates with the video's intended audience. Additionally, Elai.io allows adjustments to slide backgrounds, integration of images or shapes, and the addition of animations, offering a tailored approach to scene design. The platform also supports diverse voiceover options, with a range of AI-generated voices available or the ability to upload custom audio files. Previewing the video after editing is essential for assessing the visual coherence, voice clarity, and pacing. While the AI-generated preview helps gauge the narration, full video rendering provides a comprehensive view, enabling final adjustments before the output is finalized. Once satisfied with the content, users can initiate the rendering process, after which they receive an email notification once the video is ready for download.

To maximize the effectiveness of the video creation capabilities, several best practices are advised. Constructing clear and concise scripts improves the AI's narrative delivery and ensures that the video content is digestible for viewers. Additionally, maintaining scene durations at a reasonable length, typically under 10 seconds, can enhance viewer retention by keeping the content dynamic and engaging. Utilizing the brand kit is particularly beneficial for organizations, as it allows users to apply consistent branding elements, such as specific colors, fonts, and logos, ensuring that all videos align with the organization's visual identity. It is also important to monitor project scope in relation to the subscription plan, as each video rendering session utilizes credits. Selecting a plan that matches the frequency and duration of intended video outputs prevents unnecessary interruptions and optimizes resource management. By following these practices, users can harness the full potential of the platform to create professional, cohesive, and tailored video content suited to a variety of professional applications.

5.4 Video-to-Video Generation

Following the exploration of text-to-video models, the discussion now turns to advancements in video-to-video transformation, a domain where AI facilitates the seamless conversion of one video format into another, elevating existing video content with innovative modifications and enhancements. Video-to-video models employ sophisticated generative techniques to augment footage by enhancing resolution, transforming styles, and incorporating dynamic elements, expanding the creative potential for professionals across media production, advertising, and visual storytelling. These tools empower creators to transform aesthetic qualities, upgrade low-resolution footage to high-definition formats, and integrate realistic, animated elements into otherwise static scenes (Yang et al., 2023). This section examines prominent models within this category, including their foundational technologies, unique use cases, and the strategic value they offer across industries. As with other generative AI tools, video-to-video models reveal the adaptability of AI in managing diverse input modalities, establishing them as indispensable assets for creators and organizations aiming to refine and reimagine visual content with precision.

Among the most advanced models within this domain are NVIDIA's Vid2Vid, Google's Veo, and Adobe's Firefly Video Model. NVIDIA's Vid2Vid excels in translating semantic label maps, edge maps, and pose information into photorealistic, high-resolution video, achieving outputs up to 2048 × 1024 pixels—a feature highly beneficial to filmmakers, digital artists, and animators requiring lifelike video transformations (Zhuo et al., 2022). Google's Veo model further expands on these capabilities, providing 1080p resolution generation from various input types, including text, image, and video, allowing creators to apply cinematic effects and generate complex, stylistically coherent visuals across formats. Adobe's Firefly Video Model joins these models as a formidable AI tool that leverages Adobe's design expertise to enable nuanced, high-quality video transformations suited for advertising, social media, and personalized video production. Firefly's emphasis on user-friendly, customizable templates and integration with Adobe's Creative Cloud applications positions it as a practical choice for designers and brands seeking efficient video adaptation across multiple platforms (Padbidri, 2024). Together, these models underscore the growing scope of video-to-video transformation, allowing content creators to achieve elevated detail, stylistic consistency, and innovative visual effects. By examining the underlying technology and applications of Vid2Vid, Veo, and Firefly, this section will illustrate how video-to-video transformation is reshaping the possibilities of AI-enhanced visual production, empowering creators with adaptable, high-quality solutions for dynamic content creation.

5.4.1 Adobe Firefly Video Model

Adobe Firefly (https://www.adobe.com), initially discussed for its capabilities in AI-powered image generation, has expanded to encompass video creation with the introduction of the Adobe Firefly Video Model. This extension of Firefly's functionality enables users to generate video content from text-based prompts or images, thus transforming static concepts into dynamic video sequences. The Firefly Video Model is designed to integrate seamlessly into Adobe's Creative Cloud ecosystem, particularly benefiting video editors and content creators who work within Adobe applications like Premiere Pro. Through features such as Generative Extend, this model facilitates the expansion of video clips by adding supplemental footage, enabling users to visualize reshoots and create cohesive transitions within their projects. The Firefly Video Model exemplifies Adobe's commitment to streamlining AI-enhanced workflows, allowing creators to add sophisticated video elements without requiring extensive technical expertise or additional hardware investments (Padbidri, 2024).

The Firefly Video Model serves a range of use cases, broadening the scope of applications in media production, advertising, and social media content creation. With its ability to generate B-Roll footage, extend existing video timelines, and smooth out audio transitions, the model addresses common challenges in post-production by providing tailored, high-quality video segments that seamlessly integrate with other assets. This functionality is particularly advantageous for marketers and content creators aiming to produce polished, engaging video content on compressed timelines. Adobe's incorporation of this AI model into its Creative Cloud subscription plans reflects a strategic move to provide enhanced video capabilities to its user base without additional costs, positioning Firefly Video Model as a cost-effective solution within the existing subscription framework (Ch'ng, 2024). Additionally, Adobe emphasizes the model's adherence to legal and ethical standards, as its AI tools are trained on licensed and permitted content, addressing industry concerns related to intellectual property and commercial safety. In sum, the Adobe Firefly Video Model represents an evolution in generative AI, extending Adobe's offerings in AI-enhanced video production within a structured, commercially responsible framework designed to meet the demands of modern content creation. Next, a guide to using the platform shall be presented and summarized in Table 5.11 and can be found on the official site (https://www.adobe.com/learn/firefly).

Adobe Firefly's Video Model offers a robust, AI-driven solution for generating high-quality video content, merging accessibility with creative control. Users begin by accessing Adobe Firefly through their web browser and signing into their Adobe accounts, which enables seamless entry to Firefly's video generation features. After logging in, users select the "Generate Video" option from the main interface, initiating the creation process. The platform supports the generation of dynamic video sequences from either text prompts or an optional reference image, with each input contributing to the stylistic and narrative structure of the output. Crafting the text prompt with precision is essential; users are advised to detail visual elements such as lighting, cinematography style, and aesthetic mood to guide the AI effectively. If a reference image is included, it will serve as the initial frame, providing a visual anchor that helps align the generated content with the creator's vision.

Within the settings, a range of customization options allows users to refine the technical and creative attributes of the video. Configuring aspect ratio and frame rate ensures the output aligns with the intended platform or project specifications, while the selection of camera settings—such as shot size and angle—adds a layer of professional visual control. Notably, the tool provides several camera motion options, from zooms to tilts and handheld effects, allowing users to introduce fluidity and directionality into their videos. For advanced users seeking consistency across multiple generations, Adobe Firefly includes an optional seed number function, which governs the randomness in the creative output. Utilizing a consistent seed number with identical prompts allows for the recreation of similar videos, a feature particularly useful for iterative design processes.

Table 5.11 Guide to generating videos using Adobe Firefly.

Step	Action	Details
1. Access Adobe Firefly	Navigate to the Adobe Firefly website.	Open your preferred web browser and go to Adobe Firefly. Ensure you are using the Chrome browser, as it is currently the supported platform for video generation.
2. Sign In	Log in to your Adobe account.	Click on the "Sign In" button and enter your Adobe credentials. If you do not have an account, you can create one by following the on-screen instructions.
3. Select 'Generate Video'	Initiate the video generation process.	On the Adobe Firefly homepage, locate and select the "Generate Video" option to access the video creation interface.
4. Enter Text Prompt	Provide a descriptive text prompt.	In the "Prompt" field, input a detailed description of the video you wish to create. Be specific about elements such as lighting, cinematography, color grade, mood, and aesthetic style to guide the AI effectively.
5. Upload Reference Image (Optional)	Add an image to guide video generation.	If desired, use the "Upload" option in the Image section to provide a reference image. This image will serve as the first frame and help align the generated video with your vision.
6. Configure General Settings	Adjust aspect ratio and frame rate.	In the "General settings" section, set the desired aspect ratio and frames per second (FPS) to match your project's requirements.
7. Set Camera Parameters	Define shot size and camera angle.	Under the "Camera" section, choose the appropriate shot size (e.g., close-up, medium shot) and camera angle (e.g., eye level, high angle) to achieve the desired framing and perspective. Note that if an image is uploaded, these options may be disabled, as the camera settings will be derived from the image.
8. Select Camera Motion	Determine camera movement.	Choose from available camera motions such as zoom in, zoom out, move left, move right, tilt up, tilt down, static, or handheld to guide the camera's movement within the generated video.
9. Input Seed Number (Optional)	Control randomness in generation.	In the "Advanced" section, enter a seed number to influence the randomness of the AI's creation. Using the same seed with identical prompts and settings can help regenerate similar video clips.
10. Generate Video	Initiate video creation.	After configuring all settings, click the "Generate" button. The AI will process your inputs and generate the video. Keep the browser tab open during this process to ensure successful generation.
11. Review and Download	Access and save the generated video.	Once generation is complete, the video will appear in the queue as ready for download. Use the "History" option to access previously generated videos for comparison, iteration, and download.

Several best practices can enhance the effectiveness of Adobe Firefly's Video Model in producing visually cohesive and conceptually aligned outputs. Ensuring prompts are at least eight words in length, but not exceeding 1,800 characters, provides the AI with sufficient context to generate detailed visuals. For optimal functionality, users should work within Google Chrome on a desktop, as this environment is best suited to Firefly's processing requirements. Reviewing the generated video upon completion, accessible in Firefly's "History" section, allows users to assess the AI's interpretation and download satisfactory outputs. Additional edits or iterations may be made based on preview results, especially in complex projects where precise visual and narrative alignment is critical. By combining thoughtful prompt construction, strategic use of the customization features, and adherence to best practices, users can maximize the potential of the model as a powerful tool for producing high-quality, AI-enhanced video content that aligns with their creative objectives and technical specifications.

5.4.2 NVIDIA Vid2Vid

NVIDIA's Vid2Vid (https://www.nvidia.com/en-us/on-demand/session/gtcsj20-s21142/) represents a sophisticated AI framework engineered for high-resolution, photorealistic video-to-video transformation, developed by the research team at NVIDIA to leverage the potential of GANs. This framework facilitates the translation of structured data inputs—such as semantic label maps, edge maps, or pose estimations—into realistic video outputs. The underlying technology is optimized to generate high-quality, coherent visuals that maintain temporal consistency across frames, making it highly suitable for applications requiring smooth and lifelike animation. By translating various forms of visual data into dynamic video sequences, Vid2Vid allows users to generate realistic and immersive content that can be applied in diverse contexts, from simulation environments to entertainment production, underscoring the commitment to advancing generative capabilities within visual computing (Kumar and Singh, 2023).

The common use cases for the system span multiple industries, reflecting its versatility and effectiveness in producing photorealistic content. In autonomous driving simulations, Vid2Vid enables the creation of lifelike street scenes from segmentation maps, supporting the development and testing of self-driving technologies within controlled environments. Within the entertainment and media sectors, the framework is employed to generate realistic human motion videos and animated characters by converting pose estimations or edge maps into coherent visual sequences, a feature beneficial for character animation and special effects. Additionally, the ability to synthesize talking-head videos expands its applications in virtual conferencing, training, and education by facilitating interactive visual content creation. While NVIDIA offers Vid2Vid as an open-source project on GitHub, enabling free access for researchers and developers, the framework requires substantial computational resources, particularly high-performance GPUs, which may present additional implementation costs (Shi et al., 2024). Organizations considering the deployment of the tool should assess the necessary hardware and infrastructure to maximize the capabilities of the framework and achieve optimal video quality in alignment with NVIDIA's state-of-the-art visual processing standards.

A guide for how to use the open-source model can be seen in Table 5.12 and here (https://www.run.ai/guides/ai-open-source-projects/nvidia-vid2vid).

Utilizing NVIDIA's Vid2Vid framework for video generation is a systematic process that demands careful setup, precise configuration, and ongoing refinement to achieve optimal results. To begin, users must ensure that their system meets the technical requirements for the framework: Vid2Vid operates on Linux or macOS systems equipped with NVIDIA GPUs that support CUDA and cuDNN, providing the necessary computational power for video processing. Once the system is confirmed to be compatible, users install essential Python libraries, such as `dominate` and

Table 5.12 Guide to generating videos using NVIDIA Vid2Vid.

Step	Action	Details
1. System Preparation	Ensure system compatibility.	Verify that your system runs on Linux or macOS, has Python 3 installed, and is equipped with an NVIDIA GPU supporting CUDA and cuDNN. GitHub
2. Install Dependencies	Set up necessary libraries.	Install Python libraries such as dominate and requests using pip: pip install dominate requests. For face datasets, install dlib: pip install dlib. For pose datasets, install DensePose and/or OpenPose.
3. Clone the Repository	Obtain the Vid2Vid codebase.	Execute git clone https://github.com/NVIDIA/vid2vid in your terminal to clone the Vid2Vid repository. Navigate into the directory with cd vid2vid.
4. Download Example Datasets	Acquire sample data for testing.	Run python scripts/download_datasets.py to download example datasets provided by NVIDIA.
5. Compile FlowNet2	Prepare the optical flow network.	Execute python scripts/download_flownet2.py to compile a snapshot of FlowNet2, essential for motion estimation.
6. Download Pre-trained Models	Obtain models for specific tasks.	For Cityscapes, run python scripts/street/download_models.py. For face datasets, execute python scripts/face/download_models.py.
7. Test the Model	Generate videos using pre-trained models.	For Cityscapes at 2048x1024 resolution, run: python test.py --name label2city_2048 --label_nc 35 --loadSize 2048 --n_scales_spatial 3 --use_instance --fg --use_single_G. For face datasets at 512x512 resolution, execute: python test.py --name edge2face_512 --dataroot datasets/face/ --dataset_mode face --input_nc 15 --loadSize 512 --use_single_G.
8. Review Results	Analyze generated videos.	Access the generated videos in the ./results/ directory corresponding to your test. Review the outputs to assess quality and coherence.
9. Train with Custom Data (Optional)	Train models on your datasets.	Prepare your dataset following the structure of example datasets. Modify training scripts as needed and execute them to train models on your data.
10. Fine-Tune and Iterate	Refine models for optimal performance.	Adjust hyperparameters, experiment with different settings, and retrain models to achieve desired video quality and performance.

`requests`, through pip, followed by any dataset-specific libraries, such as `dlib` for facial datasets or DensePose and OpenPose for pose estimation tasks. With dependencies in place, users clone the Vid2Vid GitHub repository and proceed to download the example datasets provided by NVIDIA, which allow for initial testing and exploration of the framework's capabilities.

Once the foundational setup is complete, users download and compile FlowNet2, an optical flow network crucial for motion estimation within the generated video sequences. Following this, pre-trained models can be downloaded for specific tasks, such as generating photorealistic street scenes or lifelike human faces, tailored to different applications and resolutions. Testing the model involves running pre-configured scripts tailored to each dataset type—whether Cityscapes or facial data—enabling the user to generate videos with realistic details and temporal coherence. The generated video outputs are stored in the ./`results`/ directory, where they can be reviewed for visual quality and fidelity. Users aiming to expand Vid2Vid's applications may choose to train models with custom datasets, which requires adapting the data structure to match the example datasets provided, followed by script modifications and execution for training on new data. For optimal results, this process may involve experimenting with hyperparameters and retraining the model to fine-tune its performance, particularly when working with high-resolution video inputs.

Best practices for using Vid2Vid effectively include a combination of thoughtful system preparation, detailed data management, and iterative parameter tuning. Given the substantial computational demands of Vid2Vid, users are advised to utilize NVIDIA GPUs with adequate memory to manage high-resolution video tasks efficiently. Custom data preparation should involve generating accurate semantic maps, instance maps, or pose estimations to support specific transformation goals, ensuring that the data aligns closely with the intended application. Adjusting parameters, such as `--n_frames_total`, `--n_scales_spatial`, and `--n_gpus_gen`, allows users to balance video quality with processing efficiency based on their hardware resources, a practice especially valuable for those generating complex or extended video sequences. By integrating these considerations with a structured approach to Vid2Vid's setup and usage, users can maximize the framework's potential to produce visually cohesive, photorealistic videos suited to a variety of professional and creative contexts.

5.4.3 Google Veo

Google Veo (https://deepmind.google/technologies/veo/), unveiled at Google I/O 2024, represents an innovative leap in AI-powered video generation technology developed by Google DeepMind. This advanced generative model is designed to produce high-definition, 1080 p videos exceeding one minute in length, translating diverse input types—text, images, and video—into dynamic and realistic video sequences. Veo's underlying architecture enables it to generate visually rich, coherent content with minimal manual input, marking a significant advancement in automated media creation. By interpreting detailed prompts with precision, Veo provides users with the flexibility to create cinematic-quality videos that align closely with specific creative visions, making it an influential tool for professionals in media production and visual communication (Wan et al., 2024).

The primary applications of Google Veo span a range of sectors, highlighting its versatility in professional and creative domains. Filmmakers and content creators, for example, can utilize Veo to produce cinematic sequences, experiment with diverse visual aesthetics, and streamline production workflows by minimizing the need for extensive post-production editing. Within the advertising industry, Veo facilitates the rapid creation of customized promotional videos tailored to specific campaign themes, enhancing marketers' ability to engage audiences effectively. Additionally, in educational and communication settings, Veo enables the development of illustrative videos that simplify complex information, contributing to more impactful learning experiences. While Veo is currently available in a limited testing phase through platforms like Google's VideoFX, detailed pricing information has not yet been released. Select users can join the waitlist to access Veo's experimental tools, with further availability and pricing structures expected as the model progresses beyond testing (Chen et al., 2024). A guide for how to use the open-source model can be seen in Table 5.13.

Table 5.13 Guide to generating videos using Google Veo.

Step	Action	Details
1. Access VideoFX	Navigate to the VideoFX platform.	Open your web browser and go to VideoFX. Ensure you have a Google account to sign in. Google Blog
2. Join the Waitlist	Request access to Veo.	On the VideoFX page, locate the option to join the waitlist for Veo. Submit your request and await confirmation from Google.
3. Receive Access Confirmation	Await approval.	Once approved, you will receive an email notification granting access to Veo's features within VideoFX.
4. Log In to VideoFX	Sign in to your account.	Use your Google credentials to log in to VideoFX and access the Veo interface.
5. Initiate Video Creation	Start a new project.	Click on the "Create New Video" button to begin the video generation process.
6. Input Text Prompt	Describe your desired video.	In the prompt field, enter a detailed description of the video you wish to create, specifying elements like scene, style, and duration.
7. Upload Reference Media (Optional)	Provide images or videos.	If desired, upload reference images or videos to guide Veo in generating content that aligns with your vision.
8. Configure Settings	Adjust video parameters.	Set preferences such as resolution (up to 1080p), aspect ratio, and length of the video.
9. Generate Video	Initiate the creation process.	Click the "Generate" button to allow Veo to process your inputs and create the video.
10. Review and Edit	Assess the output.	Once generation is complete, preview the video. Use available editing tools to make adjustments as needed.
11. Download or Share	Finalize your video.	After finalizing edits, download the video to your device or share it directly via integrate

Using Google Veo through the VideoFX platform offers an efficient, AI-driven approach to creating high-quality videos based on text or multimedia prompts. Access to Veo requires joining the waitlist on VideoFX, Google's experimental platform, after which users receive confirmation granting them access to Veo's features. Once approved, users log in with their Google credentials and can begin a new project by selecting the "Create New Video" option. At this stage, creators enter a detailed text prompt in the input field to guide the AI's generation, describing specific aspects of the video such as setting, style, color schemes, and overall mood. If desired, users can enhance the model's accuracy by uploading reference media, including images or existing video clips, that help the AI align its output with the creator's vision. This optional input serves as a visual anchor, allowing Veo to interpret and replicate stylistic elements with greater fidelity.

Within the VideoFX environment, users also configure various technical settings, such as resolution (up to 1080 p), aspect ratio, and video length, which ensure that the final video meets both aesthetic and functional requirements. Once these parameters are set, the "Generate" function prompts Veo to process the input and render a video output. After the video is generated, creators can preview and evaluate the AI's interpretation, with options to refine the result through built-in editing tools. Adjusting elements like transitions, color grading, or pacing helps tailor the content to precise standards before downloading or sharing it directly through integrated platforms. By including an editing phase, Veo supports a flexible, iterative approach to video creation, enabling users to fine-tune results for professional-quality output.

Several best practices optimize the effectiveness of Veo and ensure that generated videos align closely with user expectations. First, specificity in the initial prompt is essential; detailed descriptions enhance the AI's ability to produce coherent and visually accurate content. Including visual descriptors—such as lighting, camera angles, or mood—ensures that Veo generates videos that closely match the intended aesthetic. Additionally, uploading reference media provides the AI with stylistic guidance, resulting in outputs that are more consistent with user-defined themes. For complex projects, using VideoFX's editing tools allows for fine adjustments, improving the video's flow and coherence. As VideoFX is currently optimized for Google Chrome, users should access it within this browser for optimal performance. Adopting these strategies, in conjunction with Veo's versatile features, equips users to produce refined, AI-enhanced videos that cater to a broad spectrum of creative and professional applications.

In examining the capabilities and methodologies across platforms like Adobe Firefly, NVIDIA Vid2Vid, and Google Veo, it becomes evident that AI-driven video generation tools offer unprecedented flexibility and efficiency for a wide array of creative applications. These tools not only streamline the production process but also enhance the ability of creators to achieve high-quality, professional results with minimal manual intervention. Each model provides unique strengths: Firefly's seamless integration with Adobe's Creative Cloud facilitates rapid content creation for designers, Vid2Vid excels in photorealistic transformations critical to simulation and animation, and Veo's adaptable inputs empower users to generate cinematic-quality content from diverse media types. Together, these platforms represent a transformative shift in video production, allowing creators to generate, iterate, and refine visual content with ease. However, as with any production workflow, the quality and impact of AI-generated videos can be further elevated through thoughtful postproduction. The following section will explore key postproduction considerations, including editing, color correction, and effects integration, that ensure AI-generated videos align seamlessly with professional standards and viewer expectations.

5.5 Post-Production Techniques for AI-Generated Video

This section addresses the essential components of post-production for AI-generated video, focusing on the tools and techniques required to transform raw AI outputs into polished, professional-quality content. As AI video generation becomes more accessible, the importance of refining these outputs through established editing workflows grows. Software such as Adobe Premiere Pro and Cinema 4D offers advanced capabilities for enhancing AI-generated visuals, enabling users to integrate transitions, apply color grading, and add depth through lighting and effects. By utilizing these industry-standard programs, editors can create cohesive visual narratives that align seamlessly with conventional footage. Throughout this section, best practices for optimizing these workflows will be presented, from setting up efficient editing pipelines to refining aesthetic details that ensure consistency and visual quality (Rahman and Ali, 2024).

Beyond traditional editing, this section will explore specialized tools that enhance both audio and visual elements of AI-generated videos, with particular attention to lip-syncing, voice synthesis, and scoring. Dialogue and lip-sync tools like Descript and DeepBrain AI are invaluable for aligning spoken audio with realistic character movements, while AI-driven scoring tools like AIVA and Soundraw provide dynamic soundscapes that enhance narrative depth. Each tool is discussed with recommendations on how best to incorporate them into a cohesive post-production workflow, allowing creators to maximize the impact of AI-generated visuals through thoughtful integration of dialogue, music, and effects (Kizer, 2024). Leveraging these tools and adhering to structured workflows allows content creators to elevate their projects, transforming initial AI outputs into compelling, fully realized video productions.

5.5.1 Editing and Enhancing in Professional Video Software

To integrate AI-generated content effectively into professional video editing software such as Adobe Premiere Pro (https://www.adobe.com) and Cinema 4D (https://www.maxon.net), following a structured approach with specific steps enhances the quality and consistency of the final output (Table 5.14). In Adobe Premiere Pro, the process begins by importing AI-generated clips into the project panel. Here, clips should be organized according to scenes or project phases to maintain an efficient workflow. A critical first step is to align the color grading of AI-generated footage with other project elements. Using Premiere's Lumetri Color panel, editors can adjust temperature, tint, contrast, and saturation to ensure a consistent color scheme throughout the video. This

Table 5.14 Video editing in Post-Production.

Software	Step	Action
Adobe Premiere Pro	**Import AI-Generated Clips**	Import clips into the project panel and organize by scenes or project phases to maintain efficient workflow.
	Color Grading	Use the Lumetri Color panel to adjust temperature, tint, contrast, and saturation, ensuring consistency with other footage.
	Transition Effects	Apply transitions like crossfades and dissolves to facilitate smooth transitions between scenes, enhancing narrative flow.
Cinema 4D	**Import AI-Generated 3D Assets**	Import 3D models and adjust lighting and texture mapping to visually align with the project's aesthetic.
	Lighting Adjustments	Modify lighting for realistic shadowing and depth, especially if blending 3D animations with live-action footage.
	Depth of Field and Motion Blur	Apply depth-of-field and motion blur effects to mimic natural camera focus and motion, integrating AI-generated elements seamlessly.
Post-Production Workflow	**Establish Folder Structure**	Organize files by type (raw footage, AI assets, audio, graphics) for quick access and streamlined workflow.
	Create Rough Cut	Construct an initial sequence of clips to visualize and identify necessary adjustments for narrative cohesion.
	Fine-Tuning	Perform color corrections, synchronize audio, and add visual effects as needed for consistency and quality.
Project Review	**Frequent Previews**	Regularly preview the project to evaluate the alignment of AI-generated content with other elements and make iterative improvements.
Version Control	**Track Changes**	Implement version control using tools like project snapshots in Premiere or Git for multimedia, allowing reversion to earlier versions if needed.
Collaboration	**Cloud Storage and Feedback**	Use cloud storage for file sharing and feedback from team members, ensuring efficient collaboration and smooth integration of AI elements.

technique helps bridge the visual gap between AI-generated clips and traditionally captured footage. Transition effects such as crossfades and dissolves also play an important role, facilitating smooth scene transitions and enhancing narrative flow (Erpelding et al., 2024).

Cinema 4D, particularly well-suited for projects involving 3D modeling and animation, enables the precise integration of AI-generated 3D assets. Import the 3D models and adjust settings such as lighting and texture mapping to ensure that these elements visually align with the scene. Lighting adjustments are essential for achieving realistic shadowing and depth, especially if the AI-generated content includes 3D animations that need to blend with live-action footage. To add further realism, the depth-of-field and motion blur effects of Cinema 4D can be applied. These effects mimic natural camera focus and movement, resulting in AI-generated elements that feel cohesive within the overall visual aesthetic of the video.

Building an efficient post-production workflow tailored to AI-enhanced projects involves several best practices. Begin by establishing a clear folder structure for all assets, categorizing files by type (e.g., raw footage, AI-generated assets, audio, graphics) to facilitate quick access. Creating a rough cut at the start of the editing process helps in visualizing the sequence of clips and identifying any adjustments required to maintain narrative cohesion. Following this, a fine-tuning phase should focus on color corrections, audio synchronization, and the addition of visual effects as needed (Li, 2024). Frequent project previews are invaluable, enabling editors to assess the alignment of AI-generated content with other video elements and identify areas for improvement.

To manage iterative changes effectively, version control is recommended for tracking project edits and preserving previous versions. Version control systems, such as project snapshots within Premiere Pro or external software like Git for multimedia, support collaborative editing by maintaining a record of all changes, allowing editors to revert to earlier versions if needed. Collaboration is further enhanced through cloud storage solutions, which streamline the process of sharing files with team members and obtaining feedback. The steps outlined here represent best practices that editors can use to integrate AI-generated content seamlessly into their projects, resulting in a cohesive, high-quality video that maintains professional standards.

5.5.2 Special Effects and Visual Consistency

Achieving visual consistency in video production, particularly when incorporating AI-generated content, necessitates meticulous attention to color grading and lighting adjustments (Table 5.15). Color grading serves as the initial step in creating a cohesive visual aesthetic across all footage by modifying the color balance, contrast, and saturation. This process is instrumental in eliminating visual discrepancies across scenes and establishing a uniform mood. For instance, an outdoor scene filmed under natural light may need to be warmed to match indoor footage, ensuring continuity. Adobe Premiere Pro's Lumetri Color panel, a popular tool for color correction, enables editors to fine-tune temperature, tint, and saturation with high precision (McCauley and Swerzenski, 2024). Through its *Basic Correction* and *Creative* sections, editors can adjust exposure, contrast, and apply Look-Up Tables (LUTs) to develop a consistent aesthetic that aligns with the intended narrative. By carefully adjusting these parameters, editors can enhance the viewer's emotional engagement by aligning the visual tone with the storyline.

Lighting adjustments, closely linked with color grading, refine exposure and shadow details to contribute to the overall visual harmony of the video. By manipulating light levels in post-production, editors can create a seamless visual flow, whether they aim to soften shadows for a more balanced composition or heighten contrast for dramatic emphasis. Software such as DaVinci Resolve (https://www.descript.com) offers sophisticated controls over these aspects, enabling

Table 5.15 Achieving visual consistency in Post-Production.

Process	Step	Description
Color Grading	Adjust Color Balance	Modify color balance, contrast, and saturation to create a unified aesthetic across all footage, aligning with the desired narrative mood. For example, warm outdoor scenes to match indoor footage. Use tools such as Adobe Premiere Pro's Lumetri Color panel for adjustments.
	Use Look-Up Tables (LUTs)	Apply LUTs as a base color profile and refine using *Basic Correction* and *Creative* tools in Lumetri Color to ensure consistent color tones, enhancing viewer engagement and mood.
Lighting Adjustments	Refine Exposure and Shadows	Adjust lighting levels to achieve balanced exposure and shadow detail across scenes. Use *Lift* (shadows), *Gamma* (midtones), and *Gain* (highlights) adjustments in DaVinci Resolve for precise calibration, referencing waveform monitors for accuracy.
	Match Light Sources	Match lighting properties of AI-generated and traditional elements, ensuring no clip disrupts continuity. Use DaVinci Resolve's advanced lighting controls to bring cohesion and emphasize thematic elements.
Transitions	Apply Transitions	Use cuts, fades, and wipes to create smooth scene transitions, such as fade-to-black for shifts in pacing. Apply and adjust transitions like *Cross Dissolve* and *Dip to Black* in Adobe Premiere Pro's *Effects* panel to ensure flow that enhances viewer immersion.
Motion Effects	Add Dynamic Movement	Apply pans, zooms, and tracking effects to simulate camera motion, adding depth. Set keyframes for *Scale* and *Position* in Premiere Pro's *Effect Controls* to draw attention to thematic focal points, such as zooming in on a key object.
3D Effects	Integrate 3D Elements	Add shadows, reflections, and 3D models for spatial depth, blending 3D effects with traditional footage for realism. Use Cinema 4D or Blender to create 3D assets and apply lighting that matches scene properties to achieve a cohesive visual experience.
Structured Workflow	Rough Cut	Begin with a rough cut to establish the basic sequence of clips, visualizing the narrative structure and identifying required adjustments for scene consistency.
	Fine-Tuning Phase	Apply color grading, lighting adjustments, and motion effects in a detailed phase, performing frequent previews to ensure alignment between AI-generated and traditional footage. Each adjustment should build cohesiveness, maintaining high production standards and ensuring all elements align with the video's overall visual aesthetic.

editors to adjust *Lift* (shadows), *Gamma* (midtones), and *Gain* (highlights) with precision. Using scopes like waveform monitors or histograms allows for the meticulous calibration of light and shadow across scenes. This attention to lighting ensures that no scene stands out disproportionately, thereby supporting the continuity of the visual narrative. When applied rigorously, lighting adjustments transform footage by reinforcing thematic elements, drawing attention to focal points, and creating atmospheric depth (Wei, 2024).

The strategic use of seamless transitions and motion effects further supports a cohesive visual experience by ensuring smooth transitions between scenes and enhancing narrative flow. Transition effects such as cuts, fades, and wipes play a pivotal role in guiding the viewer's attention, allowing shifts in scene or tone without abrupt visual interruptions. For example, a fade-to-black transition may serve as a deliberate break from an action scene to a more introspective moment, signaling a change in pace and tone. Adobe Premiere Pro's *Effects* panel offers transitions such as *Cross Dissolve* and *Dip to Black*, which can be easily customized for duration and pacing. By adjusting these transitions according to the rhythm of the narrative, editors maintain an uninterrupted flow that bolsters viewer immersion, preventing abrupt breaks that could disrupt the audience's engagement with the story.

Motion effects, including pans, zooms, and tracking shots, introduce dynamic movement, elevating the visual interest and flow of the video. When used judiciously, these effects simulate camera motion in otherwise static shots, adding layers of depth and perspective that enhance the viewer's engagement. In Adobe Premiere Pro, editors can utilize the *Effect Controls* to set keyframes for properties such as *Scale* and *Position*, simulating a zoom or pan effect over a still image. For example, gradually zooming in on a significant object within the frame can draw attention to thematic symbols or focal points, creating a sense of anticipation. Such motion effects serve to heighten the visual impact of each shot, contributing to a cohesive aesthetic that aligns with the pace and rhythm of the narrative, while adding a dynamic quality that keeps the viewer visually engaged (Azzarelli et al., 2024).

Incorporating 3D effects within video production further elevates the production value, enhancing realism and spatial depth. 3D effects range from subtle enhancements, such as adding shadows or reflections, to more complex integrations, such as blending 3D models with live-action footage. For example, Cinema 4D or Blender (https://www.blender.org) provides editors with tools to design and integrate 3D effects, creating a sense of immersion that transforms the viewing experience. When 3D models are layered seamlessly with traditional footage, they create the illusion of tangible depth and space within the scene. These programs allow editors to simulate physical light properties, ensuring that shadows, reflections, and textures harmonize with the surrounding environment, achieving a seamless integration that feels cohesive within the video's aesthetic framework (Reddy et al., 2024).

Finally, ensuring visual consistency in AI-enhanced videos involves adopting a structured workflow that facilitates an organized approach to editing and post-production. Beginning with a rough cut allows editors to establish the fundamental sequence of clips, offering a preliminary view of the narrative structure and identifying areas requiring adjustment. This initial stage is followed by a fine-tuning process, where specific attention is given to color grading, lighting adjustments, and motion effects. Each of these elements is previewed frequently to evaluate their integration within the narrative, enabling incremental improvements and ensuring that AI-generated content aligns cohesively with traditional footage. By following a structured process, editors can create a professional-quality video where every visual element aligns seamlessly, enhancing the storytelling and providing a unified, immersive experience for the viewer.

5.5.3 Dialogue, Lip-Syncing, and Voice Synthesis

The integration of dialogue, lip-syncing, and voice synthesis into video production has been transformed by AI-driven platforms like Descript (https://www.descript.com) and Synthesia (https://www.synthesia.io), which enable a level of precision and efficiency previously unattainable (Table 5.16). Descript offers a robust suite of tools for editing audio and video, with features such as Overdub, an AI-driven voice cloning tool that enables creators to train a model on their own voice. The generation of lifelike voiceovers through simple text-to-speech conversion is made possible with Overdub, allowing editors to seamlessly revise narration without returning to traditional recording methods. For instance, if a segment of dialogue requires an adjustment after recording, editors can type in the new text, and Overdub will generate a replacement audio clip that is nearly indistinguishable from the original voice (Alshahrani and Maashi, 2024). This capability streamlines the production workflow, reducing the need for re-recording sessions and enabling fast revisions. Additionally, the transcription accuracy and editing capabilities of Descript contribute to efficient workflows for multimedia content creators across industries.

As noted, Synthesia, a platform specializing in AI-driven video generation, advances this functionality by providing digital avatars that can deliver content in over 140 languages. This feature is especially beneficial for multilingual projects, allowing producers to create localized versions of videos for international audiences without requiring multiple voice actors or studio resources. For exam users, Synthesia represents a cost-effective and efficient solution for organizations seeking to scale their media outreach on an international level.

Achieving realistic lip-syncing is another critical component in enhancing the authenticity of animated characters and digital presenters. Platforms such as DeepBrain AI (https://www.elevenlabs.io) and HeyGen (https://heygen.com) have made significant advancements in this area, employing sophisticated machine learning algorithms to create avatars with highly accurate facial expressions and lip-sync. DeepBrain AI, for instance, generates virtual characters whose mouth movements correspond precisely with audio input, producing natural, expressive dialogue delivery. This technology is particularly valuable in applications requiring extended interactions with digital characters, such as customer service simulations or virtual classroom environments, where the credibility of the avatar directly impacts user engagement and trust. By ensuring that lip movements align with spoken words, these tools help bridge the gap between synthetic and human elements, contributing to a smoother, more immersive viewer experience.

Table 5.16 Dialogue, Lip-synching, and Voice Synthesis in Post-Production.

Aspect	**Tool/Platform**	**Functionality**	**Application**
Voice Synthesis	Descript (Overdub)	Provides AI-driven voice cloning for realistic voiceovers. Overdub allows users to train the model on their own voice for seamless text-to-speech conversion, enabling rapid edits without re-recording.	Used to revise narration in video projects by typing new text to replace or add to recorded dialogue, saving time and resources for content creators across industries. For example, a change in corporate script can be made immediately without requiring a new recording session.
Multilingual Avatars	Synthesia	Offers over 140 language options and customizable digital avatars for multilingual content delivery, eliminating the need for live presenters or traditional recording setups.	Particularly advantageous in corporate training, marketing, and educational videos intended for international audiences. For instance, a single video can be adapted into multiple languages, enabling organizations to expand their global reach and communicate effectively with diverse audiences.
Lip-Syncing and Realistic Facial Expressions	DeepBrain AI	Uses machine learning algorithms to generate avatars with accurate lip-sync and facial expressions that correspond to audio input, enhancing the believability of virtual characters.	Valuable in customer service simulations and virtual classrooms, where maintaining the realism of avatar interactions supports engagement and user trust. For example, DeepBrain AI can create a virtual teacher whose lip-sync matches lesson narration, improving immersion for online students.
Dynamic Movements and Expression in Avatars	HeyGen	Provides avatars with subtle facial expressions, natural body language, and accurate lip-sync, aligning audio with realistic visual cues for a cohesive presentation.	Used in educational or corporate videos where avatar relatability is essential. For example, an avatar may maintain eye contact, smile, or react appropriately during a presentation, making content more engaging and accessible for viewers.
Efficient Narration and Multilingual Reach	Synthesia	AI-powered voice synthesis that enables efficient production of narration in multiple languages, supporting inclusive content creation.	Facilitates the creation of training and promotional materials that are accessible to international and multilingual audiences. This feature is particularly useful for companies and educational institutions aiming to produce localized versions of their content for global dissemination.
Enhanced Storytelling in Animation and Games	DeepBrain AI and HeyGen	Offers high-quality lip-sync and realistic facial expressions that elevate character believability in animated and gaming contexts.	Enhances immersion and storytelling quality in animated content, such as educational videos and game cutscenes. For example, characters can deliver emotionally engaging dialogue with accurate lip-sync, contributing to narrative depth and player or viewer involvement.

HeyGen extends these capabilities by offering an array of avatars that vary in style, ethnicity, and appearance, enabling creators to select or design characters that resonate with their target audience. In addition to realistic lip-syncing, HeyGen avatars exhibit subtle facial expressions and natural body movements that add an additional layer of realism to digital interactions. For instance, an educational video created with HeyGen can feature an avatar that maintains eye contact, smiles at appropriate moments, and displays a range of facial expressions, enhancing the role of the avatar as an effective stand-in for a human presenter (Brünner and Ebner, 2024). The use of adaptive facial and motion algorithms ensures that the visual representation is in harmony with the audio content, thereby maintaining the authenticity and professionalism of the production. These advanced features are critical in educational and corporate settings, where nonverbal cues contribute significantly to viewer comprehension and engagement.

The practical applications of dialogue, lip-syncing, and voice synthesis tools are extensive, covering a range of media formats and content types. For example, in educational media, voice synthesis enables rapid production of narration for instructional videos and audiobooks, minimizing the time and resources traditionally required for recording sessions. In animation and gaming, realistic lip-syncing enhances character believability, supporting more immersive storytelling and emotional resonance. Moreover, the multilingual capabilities of platforms like Synthesia facilitate content delivery to global audiences, promoting inclusivity and broader reach. By integrating AI-powered dialogue and lip-syncing tools into the production pipeline, creators can enhance both the quality and efficiency of their projects. These technologies not only streamline workflows but also enable new possibilities for interactive media, where responsive avatars and localized content are increasingly essential.

5.5.4 AI-Generated Music and Scoring

AI-generated music and scoring have transformed contemporary video production by providing accessible, high-quality options for enhancing narrative depth and emotional resonance (Table 5.17). Platforms such as AIVA (https://www.aiva.ai) and Soundraw (https://soundraw.io) exemplify these advancements by allowing creators to generate customized background music that aligns with diverse project requirements. AIVA, an AI-driven composition assistant, offers more than 250 music styles, from classical and cinematic to jazz and electronic, allowing users to tailor soundscapes to match various themes and genres. For example, a historical documentary might use AIVA to create a classical orchestral score that evokes a sense of grandeur, while a tech-focused promotional video might employ an electronic track to underscore a futuristic tone (Butler, 2024). Soundraw, on the other hand, provides an interface that enables users to generate royalty-free music and adjust parameters such as tempo, instrumentation, and mood, making it suitable for creators seeking specific stylistic variations within a single track. This capability is invaluable for creators aiming to produce unique scores that support their creative vision without the need for extensive musical expertise or traditional scoring resources (Atanacković, 2024).

Table 5.17 Music and Scoring Post-Production.

Aspect	Tool/Platform	Functionality	Application and Example
Diverse Music Styles	AIVA	Offers over 250 music styles, including classical, cinematic, jazz, and electronic, to create soundscapes that align with various genres and themes.	For a historical documentary, AIVA could generate a classical orchestral score to evoke grandeur, while an electronic track might be used in a tech-focused video to suggest a futuristic tone. These options allow creators to match the music style to the video's narrative and aesthetic needs without the need for extensive musical knowledge.
Customization and Editing	Soundraw	Provides an interface for generating royalty-free music with adjustable parameters like tempo, instrumentation, and mood, enabling fine-tuning within a single track.	In a promotional video, Soundraw could be used to adjust tempo and mood, tailoring the music to fit specific project needs. For instance, a softer track could be customized to increase in intensity during key moments, ensuring a cohesive sound that mirrors visual dynamics and supports storytelling.
Emotional Alignment	AIVA, Soundraw	Analyzes tempo, harmony, and instrumentation to create music that aligns with scene dynamics, enhancing emotional impact.	For a suspenseful scene, AIVA could create a composition with sharp crescendos and rhythmic intensity to build tension, while a reunion scene might use a gentle, melodic piece to convey warmth. This functionality helps creators use music to amplify the intended emotional tone of each scene, creating a more immersive viewer experience.
Adaptive Soundscapes	AIVA, Soundraw	Generates adaptive music that can adjust in real-time to changes in scene intensity and character actions, particularly useful for interactive media such as video games.	In an action game, AI-generated music can respond to player actions by increasing tempo and adding dissonance during challenging encounters, shifting to a triumphant melody when a challenge is completed. This adaptability creates a dynamic, interactive experience, reinforcing player engagement and the narrative flow of the gameplay.
Cultural and Multilingual Scoring	AIVA	Provides culturally specific music options, integrating traditional instruments and regional styles for projects targeting global or multicultural audiences.	For a Japanese cultural documentary, AIVA can generate music with traditional Japanese instruments, like the koto or shamisen, adding cultural authenticity. This feature enables creators to localize music in a way that resonates with diverse audiences, helping to foster inclusivity and cultural relevance in multimedia projects.
Workflow Efficiency	Soundraw, AIVA	Supports quick iteration and modification, allowing real-time adjustments to music style and tone, which streamlines production workflows and reduces reliance on traditional composers.	For a marketing video, Soundraw can generate multiple music variations with slight tonal adjustments to test which version complements the visuals best. This agility in production saves time and resources, enabling rapid adjustments based on feedback and allowing teams to produce high-quality scores efficiently.

Effective scoring hinges on the alignment of music with scene dynamics and emotional tone, which is crucial for reinforcing the narrative impact of visual media. AI music generators like AIVA and Soundraw analyze parameters such as tempo, instrumentation, and harmonic progression to generate compositions that match the emotional landscape of specific scenes. For instance, a suspenseful chase sequence might benefit from an intense, rhythmic score with a driving beat and sharp crescendos to amplify the sense of urgency, while a scene depicting a reunion between characters may call for a softer, melodic piece that conveys warmth and nostalgia. By adjusting elements such as tempo and harmony, AI-generated music can adapt to these nuanced emotional requirements, enabling creators to use sound as a tool for mood reinforcement and audience engagement. In this way, AI-driven scoring solutions elevate the audiovisual experience by ensuring that the music is not merely background noise but an integral part of the storytelling process, enhancing the emotional depth of each scene.

The adaptability of AI-generated music further enhances its utility, particularly in applications requiring real-time responses to narrative shifts. Adaptive soundscapes, which adjust to changing intensities and plot developments, are particularly advantageous for interactive media such as video games, where music must respond dynamically to player actions and environmental changes. For example, in an action game, background music might increase in tempo and introduce dissonant chords as a player approaches a challenging encounter, creating an atmosphere of tension and anticipation. If the player successfully navigates the challenge, the music might shift to a more harmonious, triumphant melody, providing auditory feedback that reinforces the sense of accomplishment. This real-time adaptability contributes to a more immersive experience, allowing players to feel more connected to the game environment as the music evolves with their journey. In narrative-driven games, AI-generated soundscapes help maintain a consistent emotional tone while dynamically adjusting to the storyline, fostering a deeper emotional connection between players and the virtual world (Dash and Agres, 2024).

AI-driven music solutions also enable creators to easily produce multilingual or culturally specific scores, supporting content that is inclusive and accessible to global audiences. By leveraging AI platforms, producers can generate music that incorporates culturally relevant instruments or styles, which is essential for projects aimed at international or multicultural markets. For example, a documentary about traditional Japanese culture might benefit from music that integrates traditional Japanese instruments, such as the koto or shamisen, to enhance cultural authenticity. AI platforms like AIVA offer libraries of regionally inspired styles and instruments, allowing creators to produce scores that resonate with target audiences without extensive knowledge of cultural musicology. This capability is particularly beneficial for global brands or educational content creators who aim to produce localized versions of their media that feel authentic and culturally attuned. In addition to meeting the aesthetic expectations of diverse audiences, culturally adaptive scores foster inclusivity by enabling viewers from different backgrounds to connect with the content on a more personal level.

Incorporating AI-generated music into the production pipeline also streamlines workflows, allowing teams to iterate quickly and efficiently while maintaining high production quality. Traditional scoring processes, which require composers, musicians, and studio time, can be time-consuming and costly, especially for projects with tight deadlines. In contrast, AI-generated music can be produced and modified in real-time, enabling editors to experiment with different musical styles and tonal adjustments without disrupting the overall production schedule. For example, a marketing team developing a product launch video can use Soundraw to generate multiple music variations, each with slight changes in tempo or instrumentation, and test which version best complements the visuals and messaging. This flexibility supports a more agile production process, where adjustments can be made quickly to align with creative feedback. By integrating AI-driven music platforms into their workflow, production teams can save time and resources, resulting in a streamlined scoring process that does not compromise on quality, creativity, or narrative coherence.

This chapter has provided a comprehensive exploration of AI-driven video creation workflows, offering readers insight into the tools and techniques that empower modern content creators to generate, refine, and enhance visual media with remarkable efficiency and creative flexibility. Beginning with an overview of AI art generation from text prompts, platforms such as DALL-E 3, Midjourney, Stable Diffusion, and Adobe Firefly showcase the potential of AI to transform textual descriptions into compelling visuals. Each tool offers distinct advantages, from stylistic diversity to high-resolution outputs, enabling creators to select platforms that align with their artistic goals. The chapter underscores the importance of understanding these foundational tools, as they serve as the initial stages in a creative pipeline that progresses from image generation to more complex, multimedia production.

The transition from still images to animated content represents a significant advancement in AI's role within digital media, as demonstrated through tools like Pika Labs, Kaiber, and Runway Gen-3. These image-to-video models allow creators to take static images and generate motion, enhancing storytelling capabilities through dynamic visual sequences. This chapter highlighted how these platforms have simplified the video creation process, making animation accessible even to those with limited technical expertise. Furthermore, AI models designed for text-to-video generation, including Synthesia, Pictory, and Elai.io, enable creators to convert written content into video narratives with synthesized dialogue and custom avatars, opening new possibilities for fields like education, marketing, and corporate communication.

Advanced video transformations, such as those facilitated by Adobe Firefly's Video Model, NVIDIA Vid2Vid, and Google Veo, mark another leap forward in AI's applicability to professional media production. These video-to-video generation models support the adaptation and enhancement of existing footage, enabling high-quality visual effects and stylistic transformations that would otherwise require extensive manual editing. Such advancements reinforce the potential of AI to serve as a powerful asset in video production workflows, allowing creators to enhance footage consistency, integrate visual effects, and explore creative directions that add depth and realism to video projects.

The chapter concluded by addressing essential post-production techniques, including editing, special effects, dialogue and lip-syncing, and music scoring. Each section provided practical guidance on how tools like Adobe Premiere Pro, Cinema 4D, and AI music generators, such as AIVA and Soundraw, contribute to refining AI-generated content and achieving professional standards. By combining these tools with strategic workflows, creators can transform initial AI outputs into polished, cohesive videos that resonate with audiences. The practical guide presented in this chapter equips readers with a thorough understanding of how AI-driven platforms and post-production processes can be leveraged to produce engaging, high-quality video content, setting a foundation for future exploration and innovation within the evolving landscape of AI-enhanced media creation.

References

Abbott, R. and Rothman, E. (2023). Disrupting creativity: Copyright law in the age of generative artificial intelligence. *Florida Law Review,* 75: 1141.

Abdikhakimov, I. (2023, June). Unraveling the copyright conundrum: exploring AI-generated content and its implications for intellectual property rights. pp. 18–32. *In*: *International Conference on Legal Sciences* (Vol. 1, No. 5).

Akhmetova, A. and Park, S. J. (2023). Comparative assessment of 6 text-to-image AI generators: a practical analysis of features and performance. 한국콘텐츠학회 *ICCC* 논문집, 391–392.

Ali, S., Ravi, P., Williams, R., DiPaola, D. and Breazeal, C. (2024, March). Constructing dreams using generative AI. pp. 23268–23275. *In*: *Proceedings of the AAAI Conference on Artificial Intelligence* (Vol. 38, No. 21).

Alshahrani, M. H. and Maashi, M. S. (2024). A systematic literature review: facial expression and lip movement synchronization of an audio Track. *IEEE Access.*

Angelova, N. (2024). The capabilities of the art-oriented artificial intelligence Adobe Firefly and its visual advantages and disadvantages. *Journal Fundamental Sciences and Applications,* 30(1): 1–10.

Atanacković, D. (2024). Artificial intelligence: duality in applications of generative AI and assistive AI in music. *INSAM Journal of Contemporary Music, Art and Technology,* (12): 12–31.

Atilla, S. (2024). Dealing with AI-generated works: lessons from the CDPA section 9 (3). *Journal of Intellectual Property Law and Practice,* 19(1): 43–54.

Azzarelli, A., Anantrasirichai, N. and Bull, D. R. (2024). Reviewing Intelligent Cinematography: AI research for camera-based video production. *arXiv preprint arXiv:2405.05039.*

Bertelsen, L. and Short, A. R. (2023, August). State of the Art: A review of AI art generation methods for rigorous design. *In*: *International Design Engineering Technical Conferences and Computers and Information in Engineering Conference* (Vol. 87318, p. V03BT03A066). American Society of Mechanical Engineers.

Brünner, B. and Ebner, M. (2024, July). Creating educational videos with an AI avatar video generator. pp. 1165–1166. *In*: *EdMedia+ Innovate Learning.* Association for the Advancement of Computing in Education (AACE).

Butler, M. (2024). *AI and Music Digital Streaming Platforms: The Effectiveness of Implementing an Ethical Identifier to Highlight AI Compositions* (Doctoral dissertation, Institute of Art, Design+ Technology).

Chen, L., Hu, H., Zhang, M., Chen, Y., Wang, Z., Li, Y. et al. (2024). OmnixR: Evaluating Omni-modality Language Models on Reasoning across Modalities. *arXiv preprint arXiv:2410.12219.*

Cho, H. K. (2024a). A study on the work process of creating AI SORA videos. *The Journal of the Convergence on Culture Technology,* 10(5): 827–832.

Cho, H. K. (2024b). A study on the current status and qualitative development of AI Midjourney 2d graphic results. *The Journal of the Convergence on Culture Technology,* 10(5): 803–808.

Cho, J., Puspitasari, F. D., Zheng, S., Zheng, J., Lee, L. H., Kim, T. H. et al. (2024). Sora as an agi world model? a complete survey on text-to-video generation. *arXiv preprint arXiv:2403.05131.*

Ch'ng, L. K. (2024). Standing on the shoulders of generative AI. pp. 1–21. *In*: *Transforming Education With Generative AI: Prompt Engineering and Synthetic Content Creation.* IGI Global.

Dash, A. and Agres, K. (2024). Ai-based affective music generation systems: A review of methods and challenges. *ACM Computing Surveys,* 56(11): 1–34.

Dehouche, N. and Dehouche, K. (2023). What's in a text-to-image prompt? The potential of stable diffusion in visual arts education. *Heliyon,* 9(6).

Díaz Redondo, R. P., Caeiro Rodríguez, M., López Escobar, J. J. and Fernández Vilas, A. (2021). Integrating micro-learning content in traditional e-learning platforms. *Multimedia Tools and Applications,* 80(2): 3121–3151.

Erpelding, C., Beck, J., Swerzenski, J. D. and Brecheisen, T. (2024). Forum on Artificial Intelligence. *Journal of Film and Video,* 76(1): 44–55.

Fan, F., Luo, C., Gao, W. and Zhan, J. (2024). Aigcbench: Comprehensive evaluation of image-to-video content generated by ai. *arXiv preprint arXiv:2401.01651.*

Farinacci, E. (2024). Towards a renewed understanding of screen and audiovisual education: a mapping of the relationship between AI and the film industry. *Scholé: rivista di educazione e studi culturali: LXII,* 1: 183–201.

Flew, T., Gray, J. E., O'Donnell, P. and Tang, W. (2024). 17 Old and new leaders in global media markets. *De Gruyter Handbook of Media Economics*, 249.

Gopalan, R. and Kalpinagarajarao, G. (2024). The AI artistry: unleashing the power of generative AI in image creation. *J Artif Intell Mach Learn & Data Sci,* 2(4): 1471–1480.

Hadi, N. and Ainy, N. S. (2024). Increasing students' understanding of conservation using learning video media based on an artificial intelligence platform. *JHSS: Journal of Humanities and Social Studies,* 8(1): 242–247.

Huang, J. (2024). The art of AI: A Human-Centered AI (HCAI) *User Study of Integrating Image-Generative Tools in Visual Art Workflows: The Case Of Adobe Firefly.*

Hutson, J. and Harper-Nichols, M. (2023). Generative AI and algorithmic art: disrupting the framing of meaning and rethinking the subject-object dilemma. *Global Journal of Computer Science and Technology: D,* 23(1).

Isler, M., Yesilbel, B. R., Santos, V. and Bacalhau, L. M. (2023, November). Usage of artificial intelligence for advertising creation for social media marketing: ChatGPT combined with Pictory and DALL· E. pp. 73–85. *In*: *International Conference on Marketing and Technologies*. Singapore: Springer Nature Singapore.

Jai, B. R. and Shih, M. F. (2024). Technology: Limited or infinite?. *Emerging Media,* 27523543241246586.

Joseph, J. (2023). Exploring the unseen: Unleashing the potential of Synthesia AI in pedagogical approaches/Ts. Jacqueline Joseph. *RISE: Catalysing Global Research Excellence,* (3): 1–6.

Kaur, G., Kaur, A. and Khurana, M. (2024, March). A survey of computational techniques for automated video creation and their evaluation. pp. 1–6. *In*: *2024 11th International Conference on Reliability, Infocom Technologies and Optimization (Trends and Future Directions)(ICRITO)*. IEEE.

Kizer, R. J. (2024). *ADR and Post-sync Dialogue: What it is and how It's Done.* CRC Press.

Kumar, L. and Singh, D. K. (2023, June). Comparative analysis of Vid2Vid and fast Vid2Vid models for video-to-video synthesis on cityscapes dataset. pp. 1–5. *In*: *2023 International Conference on Computer, Electronics & Electrical Engineering & their Applications (IC2E3)*. IEEE.

Li, K. (2024). Application of communication technology and neural network technology in film and television creativity and post-production. *International Journal of Communication Networks and Information Security,* 16(1): 228–240.

Liao, M., Lu, H., Zhang, X., Wan, F., Wang, T., Zhao, Y. et al. (2024). Evaluation of text-to-video generation models: A dynamics perspective. *arXiv preprint arXiv:2407.01094.*

Limano, F. (2023, August). Implementation of artificial intelligence based image creation technology for conceptual ideas in 3D visual modeling. pp. 1–6. *In*: *2023 International Conference on Information Management and Technology (ICIMTech)*. IEEE.

Lind, S. J. (2024). Can AI-powered avatars replace human trainers? An empirical test of synthetic humanlike spokesperson applications. *Journal of Workplace Learning.*

Lyu, Y., Zhang, H., Niu, S. and Cai, J. (2024, May). A preliminary exploration of YouTubers' use of generative-AI in content creation. pp. 1–7. *In*: *Extended Abstracts of the CHI Conference on Human Factors in Computing Systems*.

Ma, Y. (2024). *Image and Video Generative AI in Filmmaking* (Master's thesis, New York University Tandon School of Engineering).

Martínez-Rolán, X., Corbacho-Valencia, J. M. and Piñeiro-Otero, T. (2023). Use of generative AIs in the digital communication and marketing sector in Spain. pp. 101–121. *In*: *Management for Digital Transformation*. Cham: Springer International Publishing.

McCauley, B. and Swerzenski, J. D. (2024). Losing color consciousness: platformization and precarious practices in premiere pro. *The Velvet Light Trap,* (94): 52–63.

Militsyna, K. (2023). Legal framework for output based on artificial intelligence: Ukraine's place on the global search path. pp. 559–582. *In*: *Competition and Intellectual Property Law in Ukraine*. Berlin, Heidelberg: Springer Berlin Heidelberg.

Muñoz-Basols, J., Neville, C., Lafford, B. A. and Godev, C. (2023). Potentialities of applied translation for language learning in the era of artificial intelligence. *Hispania,* 106(2): 171–194.

Naji, A. K. (2024). Employing artificial intelligence techniques to make films. *Al-Academy*, 171–180.

Ni, H., Shi, C., Li, K., Huang, S. X. and Min, M. R. (2023). Conditional image-to-video generation with latent flow diffusion models. pp. 18444–18455. *In*: *Proceedings of the IEEE/CVF Conference on Computer Vision and Pattern Recognition.*

Noetel, M., Griffith, S., Delaney, O., Sanders, T., Parker, P., del Pozo Cruz, B. et al. (2021). Video improves learning in higher education: A systematic review. *Review of Educational Research,* 91(2): 204–236.

Onyejelem, T. E. and Aondover, E. M. (2024). Digital generative multimedia tool theory (DGMTT): a theoretical postulation in the era of artificial intelligence. *Advanced Machine Learning Art Inte,* 5(2): 01–09.

Padbidri, M. (2024). *AI in Cinema: the Variations of Steamboat Willie* (Master's thesis, New York University Tandon School of Engineering).

Putjorn, T. and Putjorn, P. (2023, October). Augmented imagination: exploring generative AI from the perspectives of young learners. pp. 353–358. *In*: *2023 15th International Conference on Information Technology and Electrical Engineering (ICITEE)*. IEEE.

Rahman, M. D. and Ali, M. A. (2024). AI in video production: from script to screen. *Media and Al: Navigating*, 49.

Reddy, V. S., Kathiravan, M. and Reddy, V. L. (2024). Revolutionizing animation: unleashing the power of artificial intelligence for cutting-edge visual effects in films. *Soft Computing,* 28(1): 749–763.

Sahoo, P., Singh, A. K., Saha, S., Jain, V., Mondal, S. and Chadha, A. (2024). A systematic survey of prompt engineering in large language models: Techniques and applications. *arXiv preprint arXiv:2402.07927.*

Shi, X., Huang, Z., Wang, F. Y., Bian, W., Li, D., Zhang, Y. et al. (2024, July). Motion-i2v: Consistent and controllable image-to-video generation with explicit motion modeling. pp. 1–11. *In*: *ACM SIGGRAPH 2024 Conference Papers*.

Tsidylo, I. M. and Sena, C. E. (2023). Artificial intelligence as a methodological innovation in the training of future designers: Midjourney tools. *Information Technologies and Learning Tools,* 97(5): 203.

Wan, H., Zhang, J., Chen, Y., Xu, W. and Feng, F. (2024). Generative AI Application for Building Industry. *arXiv preprint arXiv:2410.01098.*

Wang, H., Smith, D. and Kudelska, M. (2024, August). 10x future of filmmaking empowered by AIGC. pp. 68–74. *In*: *2024 IEEE 7th International Conference on Multimedia Information Processing and Retrieval (MIPR)*. IEEE.

Wang, X., Zhang, S., Yuan, H., Qing, Z., Gong, B., Zhang, Y. et al. (2024). A recipe for scaling up text-to-video generation with text-free videos. pp. 6572–6582. *In*: *Proceedings of the IEEE/CVF Conference on Computer Vision and Pattern Recognition.*

Watiktinnakorn, C., Seesai, J. and Kerdvibulvech, C. (2023). Blurring the lines: how AI is redefining artistic ownership and copyright. *Discover Artificial Intelligence,* 3(1): 37.

Wei, D. (2024). Construction of a digital color grading laboratory based on DaVinci resolve. *Journal of Computer Technology and Electronic Research,* 1(1).

Yang, S., Zhou, Y., Liu, Z. and Loy, C. C. (2023, December). Rerender a video: Zero-shot text-guided video-to-video translation. pp. 1–11. *In*: *SIGGRAPH Asia 2023 Conference Papers*.

Zhang, Y., Kang, Y., Zhang, Z., Ding, X., Zhao, S. and Yue, X. (2024). InteractiveVideo: User-Centric Controllable Video Generation with Synergistic Multimodal Instructions. *arXiv preprint arXiv:2402.03040.*

Zhou, E. and Lee, D. (2024). Generative AI, human creativity, and art. *Eric Zhou, Dokyun Lee, Generative Artificial Intelligence, Human Creativity, and Art, PNAS Nexus,* 3(3).

Zhuo, L., Wang, G., Li, S., Wu, W. and Liu, Z. (2022, October). Fast-vid2vid: Spatial-temporal compression for video-to-video synthesis. pp. 289–305. *In*: *European Conference on Computer Vision*. Cham: Springer Nature Switzerland.

Chapter 6

The Future of AI in Film and Visual Culture

This concluding chapter examines the transformative impact of artificial intelligence on the filmmaking industry, highlighting both technological advancements and ethical considerations. The first section explores AI-driven innovations in the filmmaking process, from text-to-video generation to interactive and adaptive AI technologies, which enable new levels of storytelling and efficiency. These tools democratize filmmaking by allowing individual creators to produce feature-length films without large teams, fundamentally altering industry norms. The second section analyzes the shifting roles and skill requirements for actors, crew, and technical professionals as AI-generated characters and advanced visual effects redefine traditional job descriptions. This segment also addresses the ethical complexities of likeness rights, copyright, and the balance between human involvement and AI augmentation. Concluding with an exploration of AI's cultural and ethical dimensions, the chapter reflects on the role of the technology as both a powerful creative tool and a mirror to societal values and biases. By preserving the human touch in storytelling and developing ethical standards, the film industry can leverage AI to expand creative boundaries while safeguarding authenticity. This chapter offers a comprehensive view of the opportunities and challenges posed by AI, advocating for a balanced approach that celebrates both technological progress and the enduring art of cinema.

6.1 AI-Driven Innovation in Filmmaking Processes

The emergence of artificial intelligence in filmmaking has heralded a profound transformation, reshaping not only production workflows but also the creative potential available to filmmakers (Sun, 2024). This shift, marked by the advent of platforms such as Runway Gen-3 and Pictory, has brought once high-budget capabilities into the hands of independent creators. Text-to-video generation, for instance, allows filmmakers to envision and execute short-form productions independently (Naji, 2024). Through harnessing these tools, creators gain access to sophisticated visual effects, adaptive storytelling, and intricate animation, setting the stage for a new era of cinematic democratization where individual voices can flourish with reduced barriers to entry (Torun, 2025).

AI-driven innovation has further revolutionized the filmmaking workflow, streamlining processes from concept to post-production. As evidenced through tools like DALL-E and Midjourney, AI initiates visual storytelling at the conceptual level, enabling creators to generate preliminary visuals and storyboards. These are subsequently refined with video tools that can produce seamless transitions, complex scenes, and customized character interactions. This multi-layered AI integration brings unprecedented efficiency to the industry, allowing creators to reimagine workflows that previously relied on specialized teams. Through automation and intuitive interfaces, production timelines are significantly reduced, transforming resource-intensive tasks and expanding the creative agency of filmmakers (Rahman and Ali, 2024).

Real-world applications illustrate the depth of this transformation. Platforms like Pika Labs and Kaiber allow for dynamic video sequences from textual input, demonstrating how AI bridges the gap between concept and execution. Projects such as Cadbury's AI-driven advertising campaign, which digitally recreated Bollywood star Shah Rukh Khan, and Coca-Cola's interactive generative content initiative, reveal the commercial appeal and versatility of AI in visual storytelling. Such applications not only reflect the growing influence of the generative technology but also its capacity to blend seamlessly into traditional media formats, enhancing narratives with a unique interplay of artificial and human creativity (Çavus and Yilmaz, 2024).

However, alongside this technological potential lie complex ethical dimensions, particularly surrounding authenticity, authorship, and digital rights. Deepfakes, avatars, and automated characters bring new layers of ambiguity to questions of ownership and creative control. As creators use AI to generate lifelike representations or posthumous digital performances, these advancements necessitate a careful balance between innovation and integrity (Santo, 2024). Projects that showcase this new narrative power, like Toys "R" Us's Sora-generated origin story or Coca-Cola's collaborative user interface, demonstrate both the promise and ethical challenges inherent to the technology. These examples suggest a cautious optimism, where AI expands narrative possibilities while requiring conscientious stewardship to protect creative authenticity.

As AI evolves, its trajectory points toward deeper interactivity and adaptability in storytelling. Enhanced by advances in real-time processing and emotional expressivity, future iterations of AI promise to enable even more immersive, audience-responsive narratives. This fusion of machine-driven adaptability with human creativity paves the way for a transformative era in media, where AI acts not merely as a tool but as a collaborative partner (Chandra and Rahman, 2024). The cinematic landscape will continue to broaden, accommodating diverse voices and perspectives that redefine storytelling in ways previously unimaginable.

6.2 Redefining Roles and Skills in the AI-Enhanced Film Industry

The rise of artificial intelligence in the film industry is fundamentally reshaping traditional roles and altering skill requirements across the board (Hilal, 2024). The introduction of AI-generated characters and sophisticated visual effects tools has redefined what it means to be an actor, a crew member, or a technical professional. With AI systems now capable of creating lifelike avatars, conducting realistic lip-syncing, and generating complex special effects, many traditional filmmaking tasks that once required extensive human involvement can now be automated or significantly enhanced. This technological shift has expanded the creative potential of filmmakers but has also introduced significant changes to job descriptions and skill sets across the industry (Sookhom et al., 2023).

Actors, for instance, are witnessing their roles evolve dramatically. While they remain essential for capturing emotional depth and human authenticity, their physical presence can now be substituted or augmented by AI-generated avatars. This shift is exemplified in projects like *The Mandalorian* and *The Irishman*, where advanced deepfake and de-aging technology has allowed for the creation of younger versions of characters or entirely digital performances. These technological advancements challenge the very notion of acting, as the likenesses of actors can be recreated and manipulated digitally, altering the dynamics of traditional film production. While some argue that this opens new creative possibilities, others worry about the diminishing need for physical actors in certain contexts, particularly as AI tools improve their ability to replicate human movement and expressions with ever-increasing precision (Cogley, 2024).

For technical professionals, the implications of AI extend to their evolving skill sets. Cinematographers, VFX artists, and editors, once specialized in certain areas, now need to become proficient in AI tools that can automate or assist in their workflows. Video editors, for example, who previously spent hours cutting and refining footage, can now use platforms like Runway Gen-3 to generate dynamic video content from text prompts, drastically reducing production time. Similarly, VFX artists, who once focused on crafting individual effects shots, are now required to understand how AI can enhance visual effects through real-time generation and adaptation. As AI takes on more of the technical load, these professionals must adapt by acquiring a hybrid set of skills, blending traditional artistic expertise with technological fluency (Singh et al., 2023).

The influence of these evolving tools is also felt in the realm of crew roles, where positions that once required specialized human input are now evolving into collaborative functions with AI. For example, AI-driven pre-visualization tools are transforming the role of storyboard artists and concept designers, as generative platforms like Midjourney and Stable Diffusion can produce high-quality concept art in minutes, often in a fraction of the time it would take a human artist. These tools are also enhancing costume and set design, enabling teams to explore virtual environments and experiment with different visual styles before physically constructing sets. As the role of AI expands, the traditional division of labor on film sets is being redefined, with humans and AI working in tandem to achieve results more efficiently and creatively (Naji, 2024).

At the same time, this technological advancement raises pressing ethical questions surrounding consent, copyright, and likeness rights. With the technology now able to replicate the likenesses and voices of actors, the issue of consent becomes central. The likenesses of an actor can be digitally resurrected or manipulated in ways they may not have authorized, leading to concerns about the exploitation of their images long after their careers or lives have ended. Films like *Rogue One*, which brought back deceased actors through digital means, sparked debates over the ethical implications of such practices, particularly when family members or estates are not consulted. Furthermore, issues of copyright and intellectual property arise, as the ownership of AI-generated likenesses and performances remains legally ambiguous. Who owns the rights to a character or performance created by AI using an actor's likeness? These questions challenge established legal frameworks, pushing the industry to adapt to new realities where human and machine collaboration intersects with ownership and ethical responsibility (Barron, 2023).

As AI continues to augment filmmaking processes, the balance between human involvement and machine autonomy will remain a central consideration. While AI offers unparalleled creative tools, its encroachment into traditionally human roles necessitates careful deliberation about the future of the film industry. The evolving relationship between human creativity and AI-driven technologies underscores the need for an inclusive dialogue on the ethical, legal, and social implications of these advancements. As the industry embraces AI, the role of the artist—whether human or machine—remains integral, and the challenge lies in ensuring that technological progress enhances, rather than undermines, the art of filmmaking.

6.3 Ethical, Cultural, and Creative Considerations in AI Cinema

The rapid integration of artificial intelligence into cinema invites profound ethical, cultural, and creative considerations, challenging longstanding norms surrounding authenticity and originality. As AI continues to redefine the boundaries of filmmaking, questions about the authenticity of AI-generated performances and the originality of machine-crafted visuals become central. For filmmakers, maintaining a "human touch" in storytelling is increasingly essential in preserving the emotional resonance and nuanced insights that have historically defined cinematic art. The potential for AI to enhance, rather than replace, human creativity lies at the heart of these discussions, emphasizing that AI, while powerful, remains a tool that should amplify human vision rather than diminish it (Cavdar, 2024; Kavitha, 2023).

The ethical implications of AI cinema extend to issues of misuse, particularly with the rise of deepfake technology, which can convincingly simulate real individuals. This technology's dual potential—as a creative asset and a vector for misinformation—presents both opportunities and risks. While some creators use deepfakes to enhance narratives or recreate lost performances, the technology also harbors the capacity for deception, such as in instances of manipulated media that can alter public perception. Ensuring ethical safeguards around AI use in cinema will be vital, not only in protecting individuals' rights but also in preserving public trust in media. Through setting and enforcing standards for the responsible use of AI, the industry can mitigate these risks while fostering an environment of innovation and respect for digital ethics (Prajapat and Singh, 2024).

Culturally, AI offers an unprecedented mirror through which society can examine itself. The outputs of AI models often reveal underlying human biases embedded in the data used to train them, reflecting societal patterns that influence both the creators and the algorithms. This reflective capacity can be harnessed to promote deeper cultural awareness, as films generated with AI expose and question inherited stereotypes, cultural narratives, and aesthetic norms. By critically engaging with these AI-driven reflections, filmmakers have the opportunity to foster a more nuanced understanding of human values, exploring themes of identity, diversity, and cultural expression through an innovative, albeit computational, lens (Befera, 2024).

Creatively, the role of these platforms as a collaborator in cinema has the potential to expand the concept of creativity itself. With generative models capable of producing images, scripts, and characters, the technology challenges the traditional hierarchy between human and machine creators. Rather than viewing AI as a replacement for human ingenuity, filmmakers can approach it as a co-creator that broadens the possibilities of visual storytelling. This collaborative model, where AI assists but does not overshadow, encourages a reimagining of creativity as an inclusive process—one that integrates the precision of algorithms with the intuition and subjectivity of human artists (Yen and Lee, 2024). This is even more valid of a perspective given the goal of nearly every creator interviewed for this book being the creation of a feature-length film by one person.

Looking ahead, the role of generative artificial intelligence in cinema will likely evolve as both a tool for creative exploration and a medium for ethical introspection. As AI-driven cinema gains traction, maintaining the integrity of human storytelling, upholding ethical standards, and cultivating critical awareness of the cultural impact of the technology will remain pivotal. In this way, AI cinema can transcend its technical origins, offering insights into the evolving nature of creativity and serving as a testament to the enduring importance of human agency in the digital age. Through thoughtful engagement, the film industry can embrace AI as a catalyst for innovation and reflection, shaping a future where technology and humanity intersect in the creation of meaningful art.

References

Barron, L. (2023). AI and film. pp. 89–128. *In*: *AI and Popular Culture*. Emerald Publishing Limited.

Befera, L. (2024). Mirrored intelligence or intelligent mirror?: Historical perspectives of AI-based agents in multimedia performance. *Body, Space, & Technology Journal,* 23(1): 1–20.

Cavdar, D. (2024). *Reimagining In/Visibility, Archival Practice and the Emergence of AI Cinema: An Exploration through Refik Anadol's Artworks* (Doctoral dissertation, Concordia University).

Çavus, S. and Yilmaz, M. (2024). The effect of artificial intelligence supported advertising films on students: cola-cola masterpiece commercial movie example. *Turkish Online Journal of Educational Technology-TOJET,* 23(1): 153–162.

Chandra, B. and Rahman, Z. (2024). Artificial intelligence and value co-creation: a review, conceptual framework and directions for future research. *Journal of Service Theory and Practice,* 34(1): 7–32.

Cogley, R. (2024). *Digital Humans—A New Era of Film Technology* (Doctoral dissertation, Institute of Art, Design+ Technology).

Hilal, R. (2024). Artificial intelligence systems impact on film sets in the 21st century. *International Design Journal,* 14(4): 319–326.

Kavitha, L. (2023). Copyright challenges in the artificial intelligence revolution: transforming the film industry from script to screen. *Trinity Law Review,* 4(1): 1–8.

Naji, A. K. (2024). Employing artificial intelligence techniques to make films. *Al-Academy,* 171–180.

Prajapat, S. and Singh, A. K. (2024). Exploring the evolution and impact of artificial intelligence in science fiction cinema: an overview with financial and economic context. *Economic Affairs,* 69(2): 1097–1107.

Rahman, M. D. and Ali, M. A. (2024). AI in video production: from script to screen. *Media and Al: Navigating,* 49.

Santo, W. L. (2024). Facing Identity: The Formation and Performance of Identity via Face-Based Artificial Intelligence Technologies. *arXiv preprint arXiv:2410.12148.*

Singh, H., Rastogi, A. and Kaur, K. (2023). Artificial intelligence as a tool in the visual effects and film industry. pp. 312–316. *In*: *Recent Advances in Computing Sciences*. CRC Press.

Sookhom, A., Klinthai, P., A-masiri, P. and Kerdvibulvech, C. (2023). A new study of AI artists for changing the Movie Industries. *Digital Society,* 2(3): 37.

Sun, P. (2024). A study of artificial intelligence in the production of film. *In*: *SHS Web of Conferences* (Vol. 183, p. 03004). EDP Sciences.

Torun, A. (2025). Filmmaking and video art in the digital era. pp. 237–260. *In*: *Impact of Contemporary Technology on Art and Design*. IGI Global.

Yen, A. Y. J. and Lee, M. R. (2024). Technology competition of human-AI collaboration on the film and animation creation. *International Journal of Social and Humanistic Computing,* 4(2): 131–150.

Index

T

V

Made in the USA
Las Vegas, NV
15 September 2026

adedc49f-4719-49a9-b321-3932090db31cR01